Harnessing the Potentials of New and Advanced Materials for Developing Economics

Harnessing the Potentials of New and Advanced Materials for Developing Economics

Editor
Eric Mwangi

CENTRE FOR SCIENCE & TECHNOLOGY OF THE NON-ALIGNED AND OTHER DEVELOPING COUNTRIES
(NAM S&T CENTRE)

2013
DAYA PUBLISHING HOUSE®
A Division of
ASTRAL INTERNATIONAL PVT. LTD.
New Delhi – 110 002

ISBN 9789351242116

Centre for Science and Technology of the Non-Aligned and Other Developing Countries (NAM S&T Centre)
Core-6A, 2nd Floor, India Habitat Centre, Lodhi Road,
New Delhi-110 003 (India)
Phone: +91-11-24644974, 24645134, Fax: +91-11-24644973
E-mail: namstct@gmail.com
Website: www.namstct.org

Published by : **Daya Publishing House®**
A Division of
Astral International Pvt. Ltd.
– ISO 9001:2008 Certified Company –
4760-61/23, Ansari Road, Darya Ganj
New Delhi-110 002
Ph. 011-43549197, 23278134
E-mail: info@astralint.com
Website: www.astralint.com

Laser Typesetting : **Classic Computer Services**, Delhi - 110 035

Printed at : **Thomson Press India Limited**

PRINTED IN INDIA

Foreword

Developing countries are rich in natural resources which include raw materials such as oil, gold, timber, diamonds, platinum, uranium and chrome. They have the potential to feed themselves and others but the scramble for the developing countries' riches has often led to mismanagement, environmental degradation, and in some instances, to conflict. A number of important developments are changing the dynamics related to natural resource management and exploitation in Africa. Among these is the determination of a number of Developed countries to move from a resource-dependent to a knowledge-based economy, leveraging the potential of science and technology to ensure better management and exploitation of resources.

Africa is rich in Raw Materials which include natural resources, minerals, forests, Agricultural oriented (including animal) products and by products.

For instance, in the African context, natural raw materials were used for housing, clothing, beddings, hides and skins, artworks, cookery, pottery, ornaments (necklaces, earrings, bungles etc) just as examples.

With the current status of global industrialization, the raw materials can now be processed into new raw materials and commodity and provide local use and export products through value addition, and this could greatly improve Africa's and other Developing countries' economies.

Advanced materials and associated technologies are of great significance for the industrial growth and resultant economic development of any country in any sector, be it agriculture, civil infrastructure, transportation, energy, environmental protection, manufacturing, medicine and health, IT or aerospace. Developing countries are replete with mineral resources. However, there is an emergent need to find means for value addition to this wealth by converting them to novel and advanced materials as well as exploit them in manufacturing processes by using advanced technologies.

The new technology has traversed almost all the sectors of the economy including aerospace, transportation, information technology, environmental protection, medicine and health as well as civil infrastructure. Hence any nation that does not develop its inherent capabilities and potentials in new and advanced materials stands the risk of technological marginalisation. Moreover, the decreasing amount of raw materials needed to manufacture a unit of industrial production globally is leading to a sharp fall in the demand for the conventional raw materials, resulting in a direct negative economic impact on developing exporter nations.

The developing countries need to explore the avenues for adding value to their mineral resources and also join the elite club of nations exploiting advanced materials technology for rapid industrialisation. This therefore calls for a vigorous pursuit of a coordinated programme in the production and utilisation of new and advanced materials and the attendant technological advancements for national development.

To deliberate on these issues, the Centre for Science and Technology of the Non-Aligned and Other Developing Countries (NAM S&T Centre) jointly with the Raw Materials Research and Development Council (RMRDC) of the Federal Ministry of Science and Technology Abuja, Nigeria held a 4-day International Conference on Harnessing the Potentials of New and Advanced materials for Developing Economies was held at Abuja, Nigeria between 9th-12th August 2011.

The conference participants were primarily specialists and professionals from government departments, academic and research institutions and the private sectors in developing countries with relevant background and active involvement in New and advanced materials.

The major outcome of this workshop was the development of the Abuja 2008 Recommendations on "Harnessing the Potentials of New and Advanced Materials".

Among the key decisions were: the need for natural government and regional programmes to invest more on the effective utilization of New and advance Materials; the need to promote opportunities for collaboration and networking in New and Advanced materials research and; the need for Governments to develop policies and frameworks to promote research and development of new and advanced materials.

The scientific presentations and country reports presented in this report/ proceedings provide a stepping stone/strategic platform for development of new ideas and research programmes that will make the utilization of New and Advanced Raw Materials create an impact in Socio-economic development among the Developing countries.

It is expected that all the Governments and policy makers of these proceedings will adopt some or most of the findings into their National Development programs hence making the impact of the Abuja 2008 Conference on *"Harnessing the Potentials of New and Advanced Materials"* to be noticed globally.

To conclude, we take the opportunity to appreciate the tremendous efforts from the NAM S&T Centre and RMDRC – Nigeria in organising such a prestigious event.

It is my sincere hope that these proceedings will be widely circulated globally to all the relevant stakeholders.

Prof. Crispus M. Kiamba

Permanent Secretary

Ministry of Higher Education, Science and Technology

From the Editors' Desk

New Materials and Advanced Materials for Development and Technological Innovation

According to the UN Millennium Development Goals Task force on Science, Technology and Innovation Report of 2005, Materials are identified as playing an increasingly important role in technological innovation. Research on materials is of vital importance for technological change and particularly important for developing countries in achieving the Goals. The development of low-cost building materials could boost construction of schools and shelter in developing countries and help meet the Goal of universal primary education. By providing better living environments, low-cost building materials could reduce child mortality, improve maternal health, and improve environmental sustainability. Making the benefits of new technologies available to develop countries requires formulating a strategy that ensures that they have access to new technologies such as materials science.

Knowledge of the mechanical, electronic, ionic, and nuclear properties of metals, semiconductors, polymers, ceramics and composites, and magnetic and radioactive materials is necessary to use these materials effectively in industry.

The same report stresses that, Investment in higher education and research in materials science should form part of developing countries; strategies for industrial development. Materials – both natural and man-made – are rich with properties that can be harnessed for modern technological needs. Understanding the principles underlying the properties of materials is essential for developing new materials with properties suitable for new technologies.

Most developing countries are in the tropics. The development of cells created from new materials and photo-electrochemical cells could help formulate strategies that could help them exploit renewable sources of energy. Semiconductor research can lead to the development of new generations of integrated circuits and the solid-state memories used in ICT, semiconductor lasers, light-emitting diodes, and light-detecting devices, and technologies like photolithography.

Finally, the publication emphasizes that, Newer ceramic materials, such as piezoelectric ceramics, bioceramics, and electronic and electro-optic ceramics, provide technologically important alternatives to traditional ceramic materials. Ceramic composites, ceramic coatings, ceramic films, and glass materials (including glass ceramics, glass-ceramic composites, and conducting glasses) are important materials for industry. Special purpose polymers could be used in applications such as artificial muscles and light-emitting devices. Devices using solid-state ionic materials such (such as solid electrolytes and electrode materials) form the basis for new types of batteries, fuel cells, and sensors. All these materials could contribute to economic development in developing countries.

Background to the New and Advanced Raw Materials Workshop

The NAM Centres is The Centre for Science and Technology of the Non-Aligned and Other Developing Countries (NAM S&T Centre; www.namstct.org) is an inter-governmental organization and one of its core mandates is to undertake a variety of programmes, including organisation of workshops, symposiums, meetings and training courses in areas relevant to application of Science and technology for development.

The Raw Materials Research and Development Council (RMRDC; www.rmrdc.gov.ng) is a Parastatal under the Federal Ministry of Science and Technology, Nigeria with a specific mandate is to stimulate industrial growth through the development and utilisation of locally sourced raw materials and building technological capacity necessary for processing, acquisition, exploitation and development of these raw materials.

Therefore with the increasing important role of New and Advanced materials in technological innovation as highlighted above; and on the basis of the mandates of, the NAM Centre and RMRDC, the two organizations convened the "*International Conference on Harnessing the Potentials of New and Advanced materials for Developing Economies which was held at Abuja, Nigeria between 9th-12th August 2011*".

The conference participants were primarily specialists and professionals from government departments, academic and research institutions and the private sectors in developing countries with relevant background and active involvement in New and advanced materials.

These proceedings is a compilation of the presentations and deliberations of the above event.

The presentations fell into the following categories;

1. National policies and strategies on utilization of New and advance raw materials
2. Basic Sciences Research on New materials – chemistry/physics experimental studies

3. Value addition and development of alternative new materials to existing ones
4. Utilization plant resources as polymers and for construction industry
5. Nanotechnology
6. Laser technology

The presentations from the experts reflect the potential among the developing countries to utilize New and advanced materials for a rapid socio-economic development.

It is hoped and foreseen that the outcomes and application of the scientific recommendations of these proceedings is a major achievement to illustrate how the New and advanced materials can significantly contribute to Economic development among the developing countries.

We thank all the expert contributors to these proceedings and look forward to a workshop to review the outcomes of this event in the near future and also gather more information on the scientific research going on among the developing countries.

It was my sincere pleasure to edit the proceedings and in that respect, I wish to thank all the authors for the high quality and innovative research reported in these proceedings.

I also take this opportunity on behalf of the contributors to the NAM Centre and RMRDRC for organizing and hosting such a successful and highly relevant workshop on the niche theme on "*New and Advanced materials*".

Dr. Eric Mwangi, *BVM, PhD, OGW*
Ministry of Higher Education, Science and Technology–Kenya

Introduction

There is no specific definition for New and Advanced Materials, but a material can be defined as new and advanced, if it possesses properties that are superior to the traditional ones for given requirements, such as strength, weight, fracture toughness, corrosion resistance, physical property such as conductivity, etc. It is the value addition to the product by means of processing the material and tailoring it to meet demands for specific applications.

In the present day global competitive environment, specific innovations leading to the new and advanced materials have become crucial to the technological growth and development of a country. Exploitation of such materials can increase the competitive advantage by lowering the cost of products which incorporate them. Looking into their crucial importance to the national economies, different nations all over the world are now regularly hunting for materials up-gradation with newer technologies and even import the technology from other countries through international cooperation for meeting their demand.

Most of the developing countries are highly rich in mineral resources, but compared to the development countries, it is these countries, which in particular need to explore the avenues for adding value to their resources and join the elite club of nations in rapid industrialisation. This therefore calls for a vigorous pursuit of a coordinated programme in the production and utilisation of new and advanced materials and the attendant technological advancements for national economic and social growth.

In order to deliberate on these issues, the Centre for Science and Technology of the Non-Aligned and Other Developing Countries (NAM S&T Centre) jointly with the Raw Materials Research and Development Council (RMRDC) of the Federal Ministry of Science and Technology, Nigeria organised a 4-day International Conference on 'Harnessing the Potentials of New and Advanced materials for Developing Economies' at Abuja, Nigeria during 9-12 August 2011.

The Inaugural Ceremony commenced with the National Anthem and Opening Prayer. After the welcome address by Engr. Prof. A.P. Onwualu, Director General/ CEO, RMRDC, I made the introductory remarks and gave the background of the Conference. This was followed by a Keynote Address on 'Materials Development and Exploitation: The Determining Factor for Technological Advancement' by Engr. Prof. Daniel Oray Nnamdi Obikwelu, Head, Department of Metallurgical and Materials Engineering, University of Nigeria, Nsukka. Prof. Musa Mohammad Sada, Honourable Minister of Mines and Steel Development made a brief address, which was followed by the Inaugural Address by Prof. Ita Okon Bassey Ewa, Honourable Minister of Science and Technology of Nigeria. The Session ended with the Vote of Thanks by Dr. Seb. C. Obasi, Director, Industrial Plants and Equipment Department (IPED), RMRDC, and a group photograph.

16 countries, namely, Egypt, India, Indonesia, Iran, Iraq, Kenya, Malaysia, Myanmar, Pakistan, Sudan, Syria, Tanzania, Togo, Uganda and Zambia, and the host country Nigeria participated in the Conference. The overseas participants were from Egypt [Dr. Mona Hafez Hetta, Head, Pharmacognosy Department, Faculty of Pharmacy, Beni Suef University], India [Dr. G. Padmanabham, Associate Director, International Advanced Research Centre for Powder Metallurgy and New Materials (ARCI), Hyderabad and Dr. U.S. Tandon, Scientist 'G', Council of Scientific and Industrial Research (CSIR), New Delhi], Indonesia [Dr. Kemas Ahmad Zaini Thosin, Researcher, Sciences and Technology Park, PUSPIPTEK, Serpong], Iran [Mr. Hamid Rashidi, Sigma Aerospace Company, Tehran], Iraq [Dr. Sabeeha Abdul Jabbar Beden, Scientific Researcher, Directorate of Materials, Ceramic Centre, Ministry of Science and Technology (MOST), Baghdad], Kenya [Dr. Eric Mwangi, Deputy Director Research, Ministry of Higher Education, Sciences and Technology, Nairobi], Malaysia [Dr. TENG Wan Dung, Principal Researcher, Structural Materials Programme, Advanced Materials Research Centre (AMREC), SIRIM Berhad], Myanmar [Dr. Phyu Phyu Win, Director and Head, Nanotechnology Department and Metallurgical Research and Development Centre, Department of Technical Promotion and Cooperation, Ministry of Science and Technology], Pakistan [Dr. Amir Habib, Assistant Professor, School of Chemical and Materials Engineering (SCME), National University of Sciences and Technology (NUST), Islamabad], Sudan [Mr. Nazar Shawgi A. Ahmed, Director, Materials and Electronics Research Institute, National Centre for Research, Khartoum], Syria [Prof. Khalil AZIMEH, Department of Design and Production Engineering, Faculty of Mechanical and Electrical Engineering, Damascus University], Tanzania [Dr. Leonard Yesaya Mwaikambo, Senior Lecturer, Textile and Materials Science, Department of Mechanical and Industrial Engineering, University of Dar-Es-Salaam], Togo [Dr. Amen Yawo NENONENE, Maître-assistant, Enseignant chercheur, Unité de Recherche sur les Agro ressources et la Santé Environnementale (URASE), Ecole Supérieure d'Agronomie, Lomé University], Uganda [Prof. Barnabas Nawangwe, Principal, College of Engineering, Makerere University, Kampala], and Zambia [Mr. Bruno Mwila, Technical Officer, National Institute for Scientific and Industrial Research, Lusaka].

Overall programme of the Conference was conducted at the premises of the RMRDC in seven technical sessions, respectively co-chaired by (1) Engr. Dr. S. O. Momoh, National Agency for Science and Engineering Infrastructure (NASENI),

Abuja, Nigeria and Prof. Barnabas Nawangwe, Uganda; (2) Prof. O.S. Odusanya, Sheda Science and Technology Complex, Abuja, Nigeria and Dr. G. Padmanabham, India; (3) Dr. Babaniyi Babatope, Department of Physics, Obafemi Awolowo University, Ile-Ife, Nigeria and Mr. Nazar Shawgi A. Ahmed, Sudan]; (4) Dr. D. B. Ayo, Research Director, RMRDC, Nigeria and Dr. TENG Wan Dung, Malaysia; (5) Prof. C. O. Nwajagu, Director / Chief Executive, Scientific Equipment Development Institute (SEDI), Enugu, Nigeria and Dr. U.S. Tandon, India; (6) Engr. Dr. J. Ume, Director, Materials and Energy Technology, Projects Development Institute (PRODA), Enugu, Nigeria and Dr. Eric Mwangi, Kenya; and (7) Engr. Prof. A. P. Onwualu, DG/CEO, RMRDC and myself.

17 scientific papers presented during the Conference by the foreign participants were on 'Biologically Active Waste Product from Natural Source' by Dr. Mona Hafez Hetta of Egypt; 'Smart and Functional Materials and Structures: Design, Developments and Applications' by Dr. U.S. Tandon of India; 'Application of Lasers in Manufacturing' and 'Trends in Advanced Materials Research in India' by Dr. G. Padmanabham of India: 'The Improvement of Commercial Carbon-Steel with Ni-based Alloy Coatings' by Dr. Kemas Ahmad Zaini Thosin of Indonesia; 'Thermal Conductivity Properties of Carbon Nanocone Materials to be used in Re-Entry Vehicle by Using Molecular Dynamics Simulation' by Mr. Hamid Rashidi of Iran; 'Production of Nano Silver Water Suspension using Colloidal Silver Generator' by Dr. Sabeeha Abdul Jabbar Beden of Iraq; 'Policies and Institutional Framework for the Development and Utilisation of Raw Materials (New and Advanced) in Kenya' by Dr. Eric Mwangi of Kenya; 'Ceramic Injection Moulded Mini Crucible' by Dr. TENG Wan Dung of Malaysia]; 'Synthesis of Zinc Oxide Nanoparticles by Precipitation Method and Simple Heating Method' by Dr. Phyu Phyu Win of Myanmar; 'Sol-Gel Synthesis of Nanocrystalline Anatase Titanium Dioxide(TiO_2) at Different pH Values and Temperatures' by Dr. Amir Habib of Pakistan; 'Production and Characterisation of Dry-pressed Porcelain' by Mr. Nazar Shawgi A. Ahmed of Sudan; 'Alloying of Steels with Boron Using Laser Techniques' by Mr. Khalil AZIMEH of Syria; 'Plant Resourced Feedstokes for the Polymers Industry' by Dr. Leonard Yesaya Mwaikambo of Tanzania; 'Development of Kenaf Particleboards with Binders Based on Aqueous Extracts of Some Tanin Plants of the Flora of Togo' by Dr. Amen Yawo NENONENE of Togo; 'An Assessment of the Potential for Traditional and Modern Bamboo Construction Technologies in Uganda' by Prof. Barnabas Nawangwe of Uganda; and 'Current Status of Material Research in Zambia' by Mr. Bruno Mwila of Zambia.

11 Nigerian scientific papers presented during the Conference were on: 'Green Synthesis of Nanoparticles - Part A: Bacterial Mediated Synthesis of Gold Nanoparticles by Serratia Mercescens, Bacillus Sutilis and B Megaterium' by Ms. Stella Dozie-Nwachukwu and Prof. O.S. Odusanya of Sheda Science and Technology Complex, Abuja; 'Green Synthesis of Nanoparticles - Part B: Plant Mediated Synthesis of Gold and Silver Nanoparticles' by Dr. A.I. Aliyu of Sheda Science and Technology Complex, Abuja; 'Biosynthesis of Waste-Derived Silver Nano Particles and Their Application to Bioactive Products Development' by Dr. Jubril O. Akolade of Sheda Science and Technology Complex, Abuja; 'Challenges and Opportunities of Using Composite Materials in Automobile Industry' by Prof. S. Bolaji Hassan, Department of Metallurgical and Materials Engineering, Ahmadu Bello University, Samaru, Zaria;

'Development of Natural Fibre Filled High Temperature Engineering Thermoplastic Composites' by Prof. Umaru S. Ishiaku, Department of Textile Science and Technology, Ahmadu Bello University, Samaru, Zaria; 'Investigation of Austenitising Condition on the Tempering Characteristics of Ferritic Ductile Iron' by Mr. K. A. Bello, Ahmadu Bello University, Samaru, Zaria; 'Biotechnology for the Development of New and Advanced Materials: The Nigerian Experience' by Prof. B. O. Solomon, Director General, National Biotechnology Development Agency (NABDA), Abuja; 'Nanotechnology, Advanced Materials and Manufacturing Status in Nigeria' by Engr. Dr. S. O. Momoh of NASENI; 'Development of Polymer Ceramics with Post Consumer Poly Ethylene: Terephythalate Waste Bottles' by Dr. Olugbade Ebenezer Ajibola of Department of Chemical and Polymer Engineerig, Lagos State University, Epe, Lagos State; 'Potentials of New and Advanced Materials Development in Nigeria' by Dr. T.O. Odedele of RMRDC; and 'Bagasse (Sugarcane Waste): As Low Cost Reinforcement for Advanced Metal Matrix Composites' by Dr. V. S. Aigbodion, Department of Metallurgical and Materials Engineering, Ahmadu Bello University, Samaru, Zaria.

Before the Closing Session, I made a presentation on 'The Role of the NAM S&T Centre for South – South Cooperation in Science and Technology'. A special Participants' Interactive Forum, in which the discussion was highly interactive in nature and the participants made specific suggestions and comments, was organised, which was followed by considerable deliberation and debate on generating a set of recommendations titled 'Abuja Recommendations on Harnessing the Potentials of New and Advanced Materials for Developing Economies'. The Closing Session was chaired by Dr. (Mrs.) Dere Awosika, Permanent Secretary of the Federal Ministry of Science and Technology of Nigeria, in which the Abuja Recommendations were adopted and the participants were given the Participation Certificates.

The present book is the outcome of the deliberations and the papers presented during this International Conference on Harnessing the Potentials of New and Advanced materials for Developing Economies. I gratefully acknowledge the dynamic involvement and untiring efforts of Dr. Eric Mwangi, Deputy Director Research, Ministry of Higher Education, Science and Technology, Nairobi, Kenya in technical editing of this publication. I am also indebted to Prof. Crispus M. Kiamba, Permanent Secretary, Ministry of Higher Education, Science and Technology, Nairobi, Kenya, for sparing his valuable time in writing a 'Foreword' for this book. Last but not the least, the valuable services rendered by the entire team of the NAM S&T Centre under the guidance of Mr. M. Bandyopadhyay and Dr. V.P. Kharbanda, particularly Ms. Apeksha Yadav, Ms. Isha Sehgal and Mr. Pankaj Buttan, in compiling the presented papers and giving a shape to this volume are deeply appreciated.

I hope that this publication will serve as a valuable reference material for the researchers and professionals interested in this subject area.

Prof. Arun P. Kulshreshtha

Director and Executive Head,

NAM S&T Centre

Contents

Chapter 1

Biologically Active Waste Material of *Pisum sativum* L. (Pea)

M.H. Hetta[1], G.T. Dawood[2], K.F. Taha[2], N.Z. Yassin[3], M.E. Ali[2] and O.D. El Guindi[4]*

[1]Pharmacognosy Department, Beni Suef University, Beni Suef, Egypt
[2]Phytochemistry Department,
National Organization of Drug Control and Research (NODCAR)
[3]Pharmacology Department, National Research Centre, Cairo
[4]Pharmacognosy Department, Al Azhar University, Cairo, Egypt

ABSTRACT

With the aim of utilization of waste products as biologically active natural sources, the legumes (pericarp) of *Pisum sativum* L. (Pea), Family; Leguminosae, cultivated in Egypt were studied phytochemically and biologically. The study revealed the presence of eight phenolic acids: cinnamic, chlorogenic, vanillic, coumaric, ferulic, caffeic, gallic and syringic acids; three isoflavones: daidzein, genistein and formononetin; and two major sterols in the unsaponifiable matter: stigmasterol and β-sitosterol and the major triterpene was ursolic acid. Gas Chromatography/Mass Spectrum (GC/MS) analysis of the saponifiable fraction revealed the presence of sixteen fatty acids; palmitic acid was the major component (30 per cent). The sugars content were identified as glucose, fructose, sucrose, raffinose, verbascose and stachyose. The acute toxicity of the chloroform and 80 per cent aqueous–acetone extracts, antimicrobial screening of the saponifiable fraction and the anti-inflammatory activity of the unsaponifiable matter were studied together with the antihyperglycemic and antioxidant activities of the 80 per cent aqueous–acetone extract. Results showed that both extracts were safe up to 1.2 and 4g/kg b.w. for 15 days, respectively. The saponifiable fraction showed prominent antibacterial activity, but no effects against fungi or yeast. The 80 per cent aqueous

* *Corresponding Author:* E-mail: monahetta@gmail.com

acetone extract showed potential antihyperglycemic activity at a dose of 0.125g/kg b.w. starting on the second day of treatment. It could be concluded that the pericarp of Pea, which is considered as waste material, is a bioactive natural source as antihyperglycemic, antiinflammatory, antibacterial and antioxidant agent.

Keywords: *Pisum sativum, Pericarp, Waste product, Phytochemical, Antihyperglycemic, Anti-inflammatory, Antioxidant, Antimicrobial.*

Introduction

Food industries generate large amounts of wastes and byproducts which contain biologically active compounds. The recycling or re-usage of these wastes could be of economic benefits. Among fruits, the pericarp of Pea is separated from the seeds. The latter are processed as frozen foods. The major phytochemicals in the seeds (Habib *et al.,* 1976; Hammouda *et al.,* 1962; Tiwari, *et al.,* 2009; Sabahat and Perween, 2005, Patricia *et al.,* 1998). are the carotenoids, including lutein and zeaxanthin, β-carotene, protein, chlorophyll and phenolic compounds. It contains low levels of isoflavonoids. It also contains soluble fibre and resistant starch, which may protect against bowel cancer, together with saponins, which are thought to lower cholesterol and have anti-inflammatory, antibacterial and antifungal properties (Hedges and Lister, 2006).

Since little research was found on the pericarp of *P. sativum*, the aim of this study is the utilization of this waste product from food industry and studying them phytochemically and biologically in order to isolate biologically active components.

Materials and Methods

Material

Phytochemical Study

Plant Material

Legumes of *Pisum sativum* L. were collected at flowering stage in March 2006, from El-Kanater El-Khairya, El-Kalyobia governorate, Egypt. They were freed from seeds and the pericarp was air–dried, pulverized to fine powder, weighed and kept in an amber–coloured, tightly closed glass container for the phytochemical and biological studies.

Material for Chromatography

Silica gel 60G (70–230 mesh) for column chromatography (CC), precoated thin layer chromatography (TLC) plates (silica gel 60G F254) from E. Merck (Darmstadt, Germany).

Spray Reagents

Vanillin sulphuric acid for sterols, aniline/diphenylamine/phosphoric acid for sugars, *p*-anisaldehyde sulphuric acid and natural product polyethylene glycol (NP/PEG) for isoflavones, iodine vapors for fatty acids and ferric chloride and NP/PEG for phenolic compounds.

Reference Standards

Phenolic acids, sterols, saturated and unsaturated fatty acids and isoflavones were obtained from Sigma–Aldrich, Quimica South Madrid Spain. Sugars were obtained from Merck Company, Darmstadt, Germany.

Apparatus

Qualitative GC/MS analysis was performed on a GC HP: Hewlett Packard 5890 series II plus interfaced with HP 5989B quadruple mass spectrometer Hewlett Packard, Palo Alto, USA, with EI ion source of ionization energy of 70 eV. The MS interface was heated at 280°C, the injector at 225°C. HP–5 MS capillary column (30m × 0.25 mm × 0.25m thickness) was used. Column temperature was programmed to rise from 60°C to 300°C at a rate of 3°C/min. carrier gas, Helium with constant flow rate of 1.0ml/min. HPLC used Hewlett Packard, USA, HP 1050 series pump, UV detector and auto sampler under computer control. All analyzed compounds were separated on a lichrospher RP-18, 5µm, column (250X4mm) I.D. Merck, Darmstadt, Germany. UV-vis. Spectrophotometer used labomed double beam pc 8 scanning auto cell UVD - 3000. ^{1}HNMR was recorded at 300 MHz (Oxford 300 NMR spectrometer, Faculty of Science, Cairo University).

Biological Study

Extracts

Powdered pericarp was extracted with chloroform till exhaustion. The chloroform extract was saponified with alcoholic potassium hydroxide and following the method of Eaton (Eaton, 1989). The unsaponifiable matter was used for testing the anti-inflammatory activity, and the saponifiable fraction was used for the antimicrobial screening. Another portion of powdered pericarp was extracted with 80 per cent aqueous acetone and used for testing the antihyperglycemic and antioxidant activities.

Animals

Albino mice of 20-25g; adult male *albino* rats of 100-125g for anti-inflammatory assay and 200-250g for the antihyperglycemic test; obtained from Applied Research Centre for Medicinal Plants (NODCAR), animal house. Animals were kept under the hygienic condition and fed with well-balanced diet and water *ad libitum.*

Micro-organisms

Gram positive bacteria: *Staphylococcus aureus* (ATCC 6538), *Bacillus subtulis* (ATCC 6633). Gram negative bacteria: *Esherichia coli* (ATCC 14169), *Pseudomonas aeruginosa* (ATCC 9027). Yeast: *Candida albicans* (ATCC 10231). Fungus: *Aspergillus niger.* All the microorganisms were kindly supplied from the Microbiology Department of the Applied Research Centre for Medicinal Plants (NODCAR), Cairo–Egypt. The culture was stored in refrigerator at 5°C and reactivated on the media suitable for each microorganism.

Reference drugs

Metformin (Cidophage®) 850mg, CID Company; Ascorbic acid; Indomethacin; Alloxan monohydrate and Carrageenan were obtained from NRC and DPPH obtained from NODCAR.

Methods

Phytochemical Study

Quantitative and Qualitative Investigation of Phenolic Content

The quantitative estimation of total phenolic content was carried out spectrophotometrically using Folin-Ciocalteu reagent (Athukorala *et al.*, 2006). A part of 80 per cent aqueous acetone extract of *Pisum sativum* L. (Troszynska and Ciska, 2002), was dissolved in 10ml MeOH, 0.5ml of 50 per cent Folin-Ciocalteu reagent was added to 100µl of the MeOH extract and diluted with 8ml distilled water, left for 8min and 1.5ml of 5 per cent Na_2CO_3 was added, allowed to stand for 60min, protected from light. The absorbance was measured, to calculate the concentration of the total phenolic content, with the UV-vis spectrophotometer at 750nm using MeOH as blank. All tests were performed in independent triplicates (n=3).

Qualitative estimation of the phenolic content of the 80 per cent aqueous acetone extract of *Pisum sativum* L. was performed by fractionation of 5g of the extract over silica gel column chromatography (200 g). Elution was carried out with chloroform, followed by increasing amounts of ethyl acetate in chloroform. Fractions (100ml each) were collected, monitored by TLC using $C_6H_6/CH_3COOH/H_2O$ (37:45:18) as solvent system to yield eleven fractions. Fractions (3, 4, 7 and 10) were subjected to HPLC using authentic standards and mobile phase water/acetonitrile/phosphoric acid (85:13.8:1.2). Re-equilibration period of 10 min between individual runs was adopted. Elution was carried out at room temperature with flow rate of 1ml/min and UV detection at 320nm.

Investigation of Lipid Content

The air dried powdered plant material (600g) of *Pisum sativum* L. pericarp was subjected to exhaustive extraction with 80 per cent MeOH (4X6L), the residue of the combined extracts (88g) was diluted with water and fractionally extracted with petroleum ether (4X300ml), to yield 15g of dry extract. The petroleum ether extract was saponified. Six grams of the unsaponifiable matter was chromatographed on silica gel column using petroleum ether, petroleum ether/chloroform and chloroform, with increasing polarity, as eluant. Similar fractions were pooled together after TLC examination using toluene/EtOAc (97:3) as solvent system. Six collected fractions (1-6) were subjected to GC/MS, together with authentic sterols. The saponifiable part was methylated and subjected to GC/MS together with the available authentic fatty acids. Identification of the components of the saponifiable and the unsaponifiable matters was achieved by comparison with the mass fragmentation of the standards, also with the library search. Quantitative estimation was determined by measuring the peak areas of the identified components.

Investigation of Isoflavone Content

Powdered pericarp of *P. sativum* (600g) was exhaustively extracted with 80 per cent aqueous ethanol. The combined extracts were evaporated under reduced pressure. The residue (88g) was diluted with water and fractionally extracted with ethyl acetate

(4X300ml), to yield 6g of dry weight. Five grams was chromatographed over silica gel column using hexane and hexane/acetone, with increasing polarity, as eluant. Similar fractions were pooled together after TLC examination using EtOAc/toluene/formic acid (50:50:10) as solvent system to give four fractions (A-D). Fractions B and C were purified on sephadex HP_{20} column using methanol as eluant. Similar fractions were pooled together after monitoring by TLC using the same solvent system. From the collected subfractions B3, B4 and C4, three major compounds (1, 2 and 3) were identified using UV, ^{1}HNMR spectroscopic technique. A part of the residue (0.2g) was dissolved in 10ml 80 per cent aqueous ethanol and subjected to HPLC column using 0.1 per cent acetic acid/acetonitrile (A) and 0.1 per cent acetic acid/water (B). The isoflavones content of *P. sativum* was determined by Hewlett Packard, HPLC instrument consist of a 1050 series pump, auto sampler HP 1050 series, lichrospher C18 (5µm) column, UV detector, (USA) and detection at 260 nm wavelength. Identification of components was based on comparison of the retention times of the components of the sample and the standards (Hsieh *et al.,* 2005).

Investigation of Carbohydrates Content

Powdered *P. sativum* L. pericarp (5g) was extracted with hot water and the total carbohydrates were estimated using the general phenol-sulphuric acid assay (Kundsen, 1986; (Chaplin and Kennedy, 1994)

The reagent was prepared by dissolving phenol in water (5 per cent w/v), the water extract and standard (0.2ml of each) were mixed with 0.2ml of phenol reagent, one ml of concentrated H_2SO_4 was added rapidly to the solutions and left undisturbed for 10 minutes before shaking vigorously. The absorbance of each solution was taken at 490nm after 30 minutes. The quantitative determination of total carbohydrates was performed by measuring of peak area using TLC scanner. The soluble carbohydrates were also screened by TLC using solvent system B/A/W (2:1:1) and the spots were visualized by spraying with aniline phthalate reagent, together with comparison with authentics. The quantitative determination of total carbohydrates was performed by measuring of peak area using Hewlett Packard, HP 1050series HPLC instrument consist of a 1050 series pump, auto sampler HP 1050 series, Polygosil 60-5NH2 (250x 4.6 mm) column, IR detector (HP. 1047A), (USA). The identification was accomplished by comparison of the retention times of the components of the sample and the standards.

Biological Study

Safety Margin

The effects of intraperitoneal and oral administration of the maximum soluble concentration of the chloroform and 80 per cent aqueous acetone extract on experimental animals were observed. In this respect, 1ml of each solution was used and signs of toxicity and/or mortality were recorded up to 15 days on *albino* mice (Shayne and Christopher, 1992).

Antimicrobial Activity

The antimicrobial activity of the saponifiable fraction was determined by agar diffusion method (Perez *et al.,* 1990) against some strains of Gram +ve and Gram –ve bacteria, also against yeast and fungi. The strains were tested on soybean casein digest agar for bacteria and Sabouraud dextrose agar for yeast and fungi. Sterilized paper disks were located with different amount of the sample to be tested (100, 50, 25, 12.5, 6.25, 3.125, 1.56, 0.78 and 0.39 mg/ml) and applied on the surface of agar plates. All plates were incubated at 37°C for 24 hrs. for bacteria and at 25°C for yeast and fungi to allow maximum growth of the organisms. The lowest drug concentration resulting in a clear zone of growth inhibition around the disk after conventional incubation period, known as minimum inhibitory concentration (MIC), was calculated. The antimicrobial activity of the test agents was determined by measuring the diameter of zone of inhibition expressed in mm.

Antioxidant Activity

The free radical scavenging capacity of the 80 per cent aqueous–acetone extract was evaluated by using DPPH (2,2 diphenyl-1-picrylhydrazyl) stable radical (Blois, 1985). Briefly 1ml of 0.3mM alcoholic solution of DPPH in methanol was added to 2.5ml of Sample solutions (stock solution of 80 per cent aqueous–acetone extract (2mg/ml) and its dilutions 250, 125, 50, 25, 10, 5 µg\ml in methanol), 2.5ml of standard solutions (stock solution of Ascorbic acid (2mg/ml) and its dilutions 250, 125, 50, 25, 10, 5 µg/ml in methanol) and 2.5ml of control solution (methanol). Allowed to stand for 30 min. at room temperature then measure the absorbance value at 515nm against the blank (methanol). The percentage of inhibition (radical scavenging activity) and IC50 (the amount of sample need to decrease the absorbance of DPPH by 50 per cent) were calculated.

Anti-inflammatory Activity

Acute anti-inflammatory effect was evaluated using yeast–induced rat paw edema test described by Winter (Winter *et al.*, 1962) and Indomethacin (diclofenac sodium, 25mg/kg b.w) as a standard. Thirty six male albino rats were divided into six groups and orally treated one hr. before induction of oedema. One group receiving saline and served as positive control, three groups receiving the unsaponifiable matter in doses of 122.5, 61.25, 30.6 mg/kg b.w.; one group receiving Indomethacin (25mg/kg. b.w) and one group, served as negative control. Induction of edema was done by sub-planter injection of 100µl of 1 per cent carrageenan injection in saline into the pad of right paw. The difference in hind footpad thickness was measured immediately before carrageenan injection and 1,2,3,4 hrs after carrageenan injection with a micrometer caliber (Obukowicz *et a*l., 1998). The edema was expressed as a percentage of change from the positive control group.

Antihyperglycemic Activity

The measurement of the antihyperglycemic activity was performed by using six groups each of six male albino rats, one group served as negative control, three groups receiving the extracts at three different concentrations (0.5, 0.25 and 0.125g/kg.b.w) for 4 weeks, one group of diabetic rats served as positive control without receiving any treatment and one group receiving standard Metformin (0.18mg/kg.b.w). Induction of diabetes was carried out by injecting the rats with Alloxan monohydrate (150mg/kg) by intraperitoneal route followed by an overnight fasting

(Demehenko and Tronko, 1975). Diabetes was confirmed by the determination of fasting glucose concentration on the third day post administration of Alloxan. Blood was obtained from retro–orbital plexus of vein of anaesthetized rats. Concentration of the blood glucose level was determined at zero time, after 72 hrs, 2 and 4 weeks of administration of Alloxan.

Statistical Analysis

The data were analyzed using repeated measures one way ANOVA method. For the statistical tests a P0.05 was taken as significant.

Results

Phenolic Content

Results of quantitative estimation of phenolic content, revealed the presence of 27.5 mg/g extract of total phenolic content. Results of HPLC estimation of phenolic acids contents, Table 1.1 revealed the presence of eight phenolic acids: cinnamic, chlorogenic and vanillic acids (fraction 3), coumaric and ferulic acids (fraction 4), caffeic acid (fraction 7), gallic and syringic acids (fraction 10). Gallic acid was the major (42 per cent) followed by cinnamic acid (15.6 per cent). What is the significance of this? (presence of phenolic compounds which could be the active constituents responsible for the antihyperglycemic and the antioxidant activities.)

Table 1.1: Quantitative Estimation of Phenolic Acids of *Pisum sativum* L. Pericarp using HPLC

Peak No.	*Phenolic Acid*	*Concentration (mg/g)*	*Concentration (per cent)*
1.	Gallic acid	10.8	42
2.	Cinnamic acid	3.7	15.6
3.	Vanillic acid	2.1	8
4.	Chlorogenic acid	0.25	1.3
5.	Caffeic acid	0.28	1.4
6.	Syringic acid	0.28	1.4
7.	Coumaric acid	0.91	3.5
8.	Ferulic acid	1.03	3.9

Sterols and Fatty Acids Contents

The percentage of the saponifiable and the unsaponifiable matters of the chloroform extract were 17.5 per cent and 12.5 per cent w/w respectively. Results of fatty acids components, Table 1.2, revealed the presence of 16 fatty acids representing 96.5 per cent. The saturated fatty acids predominate and represented (64.75 per cent) while the unsaturated FA represented only 31.75 per cent. Palmitic acid was the major saturated FA (30 per cent) and oleic acid was the major unsaturated FA (19 per cent). The total USM, Table 1.3 containing 20 components representing 96.54 per cent among which eleven hydrocarbons, six sterols and three triterpenes representing 25.69 per cent, 60.1 per cent and 10.75 per cent respectively. The major sterol was stigmasterol (26.1 per cent) followed by β–sitosterol (14.25 per cent) and the major hydrocarbon was phytol (7.1 per cent), while the major triterpene is ursolic acid methyl ester (3.9 per cent). What is the significance of this? (presence of sterols and fatty acids which could be the active constituents responsible for the antibacterial activity of the pericarp)

Table 1.2: Analysis of Fatty Acid Methyl Esters of *Pisum sativum* L. Pericarp Using GC/MS

Peak No.	*Fatty Acid Methyl Ester (FAME)*	*Rt (min)*	*Concentration (per cent)*
1.	Lauric acid ($C_{12:0}$)	11.2	2.50
2.	Tridecanoic acid ($C_{13:0}$)	11	1.05
3.	Myristic acid ($C_{14:0}$)	17.4	2.00
4.	Palmitic acid ($C_{16:0}$)	19.5	30.00*
5.	Hexadecanoic acid-2-hydroxy ($C_{16:0}$)	21.6	5.50
6.	Heptadecanoic acid ($C_{17:0}$)	21.3	1.20
7.	9,12,15–octadecatrienoic acid ($C_{18:3}$)	22	0.50
8.	Linoleic acid ($C_{18:2}$)	22.5	12.25
9.	Oleic acid ($C_{18:1}$)	22.7	19.00*
10.	Stearic acid ($C_{18:0}$)	33.1	7.50
11.	Nonadecanoic acid ($C_{19:0}$)	11.4	1.50
12.	Arachidic acid ($C_{20:0}$)	26.5	1.20
13.	Behenic acid ($C_{22:0}$)	27	2.50
14.	Heneicosanoic acid ($C_{21:0}$)	27.5	3.20
15.	Tricosanoic acid ($C_{23:0}$)	28	1.10
16.	Lignoceric acid ($C_{24:0}$), (tetracosanoic acid)	33.8	5.50
	Total concentration		96.50
	Total saturated fatty acids		64.75
	Total unsaturated fatty acids		31.75

*: Major fatty acids

Rt (min): retention time in minutes.

Isoflavones Content

Three isoflavones compounds (1, 2 and 3) were isolated and identified by different physical and spectroscopic techniques

Annex 1:

Annex 2:

Annex 3:

UV λmax (MeOH) of the three compounds are in agreement with previously published data (Augusta *et al.*, 2010).

Table 1.3: Identified Components of Unsaponifiable Matter (USM) of *Pisum sativum* L. Pericarp by GC/MS

Peak No.	*Compound*	*Rt (min)*	*Relative Concentration (per cent)*
1.	Dodecane C_{12}	20.12	1.40
2.	Phytol $C_{20:1}$	23.8	7.10*
3.	Hexadecane C_{16}	25.9	1.55
4.	Octadecane C_{18}	26.1	1.36
5.	Nonadecane C_{19}	26.7	1.85
6.	Eicosane C_{20}	27.3	2.28
7.	Heptadecane C_{17}	27.6	1.05
8.	Heneicosane C_{21}	28.9	1.03
9.	Squalene $C_{30:6}$	35.04	5.25
10.	Hexacosane C_{26}	36.2	1.89
11.	Docosane C_{22}	37.2	1.20
12.	Ergost-5-en-3-ol	38.2	1.50
13.	Ergosta-4,22-dien-3-one	38.9	1.90
14.	Cholesterol	39.8	8.30
15.	Campesterol	40.0	8.05
16.	Stigmasterol	40.3	26.1*
17.	β–sitosterol	41.04	14.25
18.	Oleanolic acid methyl ester	41.9	3.85
19.	Deoxyoleanolic acid methyl ester	42.2	3.0
20.	Ursolic acid methyl ester	43.2	3.90
Total	Hydrocarbons		25.69
	Sterols		60.10

*: Major component

Rt: Retention time.

HPLC technique used for quantitative estimation of isoflavones content of *P. sativum* L. pericarp revealed presence of formononetin, diazein and genistein as 50 per cent, 30 per cent and 5 per cent, respectively.

Sugars Content

Results of quantitative estimation of soluble carbohydrate content, revealed the presence of 0.6 g/g of water extract. HPLC analysis of carbohydrates, Table 1.4, revealed the presence of six sugars: sucrose, glucose, fructose, raffinose, verbescose and stachyose.

Table 1.4: Quantitative Estimation of Sugars Content of *Pisum sativum* Pericarp using HPLC

Peak no.	*Compound*	*Concentration (mg/g)*	*per cent*
1	Raffinose	190	35 per cent
2	Sucrose	105	18 per cent
3	Glucose	45	8 per cent
4	Fructose	81	14 per cent
5	Verbascose	21	2.50 per cent
6	Stachyose	5	0.53 per cent

Safety Margin

No signs of toxicity were observed in mice, which revealed the safety of the maximum soluble concentrations of the chloroform and the 80 per cent aqueous acetone extract (1.2 and 4g/kg b.w respectively).

Antimicrobial Activity

Results of the antimicrobial activity on selected microorganisms are presented in Table 1.5. The saponifiable fraction exhibited antibacterial activity but no activities against fungi or yeast of the selected strains. It has significant inhibitory activity against *Bacillus subtilis* (MIC=0.78mg; inhibition zone=8mm diameter), followed by descending inhibitory activities against *Escherichia coli* (12.5mg, 8mm), *Pseudomonas areuginosa* (25mg, 8mm) and least activity is against *Staphylococcus aureus* (100mg, 9mm).

Antioxidant Activity

The IC50 (µg/ml) values of 80 per cent aqueous acetone extract of *Pisum sativum* pericarp for free radical scavenging activity by DPPH was 44.6, while the Ascorbic acid was 6.5.

Anti-inflammatory Activity

The unsaponifiable matter exhibited maximum inhibition (43.2 per cent) at dose (1.2 g/kg) after 4 hrs. of drug treatment, while Indomethacin exhibited 73.2 per cent of inhibition. Results are represented in Table 1.6.

Antihyperglycemic Activity

The diabetic rats fed with 80 per cent aqueous–acetone extract exhibited after 48 hrs. of treatment a significant decrease in blood glucose level (0.43 per cent) at a dose of 0.125g/kg (1/4 of maximum soluble concentration) when compared to Metformin–treated rats (0.13 per cent reduction at 0.18mg/kg). After two weeks, the three tested groups with the extracts exhibit approximately the same significant efficacy (62 per cent, 59 per cent and 63 per cent) at the doses of (0.5g, 0.25g and 0.125g respectively) when compared to the standard (21 per cent). After four weeks of treatment, the tested samples exerted more efficacies (65 per cent, 61 per cent and 66 per cent) at the mentioned doses respectively, when compared to the standard (32 per cent) (Table 1.7).

Table 1.5: Results of Antimicrobial (Antibacterial and Antifungal) Screening of Saponifiable Fraction of *Pisum sativum* L. Pericarp in different Concentrations

Micro-organisms	*Concentration of the Extract*								
	100 mg	*50 mg*	*25 mg*	*12.5 mg*	*6.25 mg*	*3.12 mg*	*1.56 mg*	*0.781 mg*	*0.39 mg*
	Diameter of Zone of Inhibition (mm)								
Gram positive bacteria									
Bacillus subtilis (ATCC 6633)	15	14	14	13	12	11	9	8*	–
Staphylococcus aureus (ATCC 6538)	9*	–	–	–	–	–	–	–	–
Gram negative bacteria									
Escherichia coli (ATCC 14169)	12	11	11	8*	–	–	–	–	–
Pseudomonas areuginosa (ACCT 9027)	1	9	8*	–	–	–	–	–	–
Yeast									
Candida albicans (ATCC 10231)	–	–	–	–	–	–	–	–	–
Fungi									
Aspergillus niger	–	–	–	–	–	–	–	–	–

*: Minimum inhibitory concentration (MIC).

Table 1.6: Results of Anti-inflammatory Activity of Unsaponifiable Matter of *Pisum sativum* L. Pericarp in different Concentrations

Groups	*Per cent of Change in Oedema Volume (Per cent of inhibition)*			
	1 Hour	*2 Hours*	*3 Hours*	*4 Hours*
Negative Control	40.7±4.0	84.1±10.2[a]	116.5±7.9[a]	126.34±6.6[a]
Indomethacin standard	29.7±1.7 (27.1 per cent)	34.1±2.4* (59.4 per cent)	38.9±2.2* (66.6 per cent)	33.9±1.2* (73.2 per cent)
Max. soluble conc. (122.5mg/kg)	42.8±3.8	64.8±2.9[a]* (22.9 per cent)	82.6±4.7[a]* (29.1 per cent)	71.78±5.4[a]* (43.2 per cent)
Half of Max. soluble conc. (61.25mg/kg)	53.7±4.1[a]	74.6±5.9[a] (11.2 per cent)	85.7±3.4[a]* (26.4 per cent)	78.1±4.3[a]* (38.2 per cent)
Quarter of Max. soluble conc. (30.6mg/kg)	51.3±4.3[a]	69.3±4.6[a]* (17.5 per cent)	92.3±4.0[a]* (20.7 per cent)	90.6±3.1[a]* (28.3 per cent)

*: Statistically significant compared to the control group at $p<0.05$.

a: statistically significant compared to Indomethacin standard group at $p< 0.05$.

Table 1.7: Results of Antihyperglycemic Activity on Rats of 80 per cent Aqueous–Acetone Extract of *Pisum sativum* L. Pericarp in different Concentrations

Groups	*Blood glucose (mg/dl) ± S.D.*			
	Zero Time	*48 Hours*	*2 Weeks*	*4 Weeks*
Negative control	102±2.3	102.5±1.3	102.16±0.7	103.5±1.7
Positive control	189.3±1.3[a]	192±1.5[a]	194.16±1.3[a]	204.3±1.7[a]
Metformin (standard) (0.18mg/kg)	353±2.3*	307.6±1.2*	277.16±1.9*	237.16±1.7*
Per cent of reduction		0.13	0.214	0.328
Max. soluble conc. (0.5gm/kg)	312.5±2.2*[,a]	210.8±1.7*[,a]	117.3±2.2*[,a]	111.3±1.2*[,a]
Per cent of reduction		0.323	0.624	0.658
Half of Max. soluble conc. (0.25gm/kg)	252.1±1.9*[,a]	186.8±2.1*[,a]	112.16±1.7*[,a]	101.1±1.1*[,a]
Per cent of reduction		0.361	0.59	0.611
Quarter of Max. soluble conc. (0.125gm/kg)	332.5±2.1*[,a]	189.3±2.3*[,a]	122.16±1.8*[,a]	111.5±1.2*[,a]
Per cent of reduction		0.43	0.632	0.665

Discussion

In our world, in which the environment pollution increases fast, the most important effort of the factories is dismissing the wastes without causing the pollution. Many food industry wastes come out beside products and could be used as useful pharmaceutical agents. The considerable antibacterial activity of the saponifiable fraction of *P. sativum* L. pericarp, especially against gram negative strains is probably due to the presence of saturated fatty acids (Syam *et al.,* 2008; Garcia *et al.,* 1999).

Phytosterols, and β-sitosterol in particular, among the major component of unsaponifiable plant material have been demonstrated to be potent anti-inflammatory (Lippiello *et al.*, 2008; Gupta *et al.*, 1980; Zheng *et al.*, 2005). Sterol extracts from different plant sources indicate their widespread distribution, and animal tests show their anti-inflammatory potency (Deepak and Handa, 2000). Phenolics constitute an important group of natural products contributing significantly to the medicinal value of a number of plants including legumes.

"According to WHO figures, in 1985, there were an estimated 30 million people with diabetes worldwide, today, there are at least 177 million people with diabetes, a nearly six fourth, increase in just over 15 year, if nothing is done to slow down the epidemic, the number will reach 300 million within 25 years " (Borse *et al.*, 2010).

The present study emphasizes that the 80 per cent aqueous–acetone extract of *P. sativum* pericarp has a potent and safe antihyperglycemic effect. Hypoglycemic activity of some leguminous plants was studied by (Singh *et al.*, 1976) and they reported that legumes in diet could reduce glucose levels in normal rats than could a normal diet. Natural α-amylase and α-glucosidase inhibitors from food-grade plant sources offer an attractive therapeutic approach to the treatment of post-prandial hyperglycemia by decreasing glucose release from starch and delaying carbohydrate absorption in the small intestine and may have potential for use in the treatment of diabetes mellitus and for obesity (Gallaher and Schneeman, 1986; Murai *et al.*, 2002). Some polyphenols are found to be effective against the inhibitory activities of α-glucosidase and aldose reductase (Jia *et al.*, 2003).

To our knowledge, this is the first report on the phytochemical and the biological studies on the pericarp of *P. sativum* cultivated in Egypt. The preliminary investigation on the antihyperglycemic efficacy of 80 per cent aqueous–acetone of *P. sativum* pericarp will be significant to proceed further in this path for the isolation of active principles responsible for antihyperglycemic activity and further studies need to be carried out to elucidate the exact mechanism of the aqueous–acetone extract on diabetes effect.

Conclusion

The secondary metabolites of the pericarp of *P. sativum* L. (Pea) which is considered as waste material, could benefit as antimicrobial, anti-inflammatory, antioxidant and antihyperglycemic agents from natural source.

References

Athukorala, Y., Kim, K.N. and Jeon, Y.J., 2006. Antiproliferative and antioxidant properties of an enzymatic hydrolysate from brown alga, Ecklonia cava. *Food Chem Toxicol.*, 44: 1065–1074.

Augusta, C., Gerardo, P., Annalisa, M., Martina, C. and Vincenzo, B., 2010. ^{1}HNMR fingerprinting of Soybean extracts, with emphasis on identification and quantification of isoflavones. *Nutrients*, 2: 280.

Blois, M.S., 1985. Antioxidant determination by the use of a stable free radical. *Nature*, 181: 1199.

Borse, L.B., Ahirrao, R.A., Borse, S.L., Sayed, H., Patil, J.K. and Desai, S.G., 2010. Search for the antidiabetic plants: Pharmacological aspects. *Pharmacologyonline,* 1: 1023–1030.

Chaplin, M.F. and Kennedy, J.F., 1994. *Carbohydrate Analysis: A Practical Approach.* Oxford Univ. Press, Oxford, New York, Tokyo, 2nd Edn., pp. 324.

Deepak, M. and Handa, S.S., 2000. Anti-inflammatory activity and chemical composition of extracts of *Verbena officinalis. Phytother. Res.,* 14: 463–465.

Demehenko, V.M. and Tronko, M.D., 1975. Testosterone metabolism in liver of alloxan diabetic rats. *Fiziol 2H(KIEV),* 21(4): 466–469.

Eaton, D.C., 1989. *Laboratory Investigation in Organic Chemistry.* McGraw Hill Book Co., New York, St. Louis. Mexico, Montreal, New Delhi, pp. 929.

Gallaher, D. and Schneeman, B.O., 1986. Nutritional and metabolic response to plant inhibitors of digestive enzymes. *Adv. Exp. Med. Biol.,* 199: 167–184.

Garcia, M.D., Saenz, M.T., Gomez, M.A. and Fernandez, M.A., 1999. Topical antiinflammatory activity of phytosterols isolated from *Eryngium foetidum* on chronic and acute inflammation models. *Phytotherapy Research,* 13(1): 78–80.

Gupta, M.B., Nath, R., Srivastava, N., Shanker, K., Kishor, K. and Bhargava, K.P., 1980. Anti-inflammatory and antipyretic activities of beta sitosterol. *Planta Med.,* 39: 157–163.

Habib, F.G.K., Mahran, G.H., Hilal, S.H., Gabrial, G.N. and Morcos, S.R., 1976. Phytochemical and nutritional studies on pigeon pea and kidney bean cultivated in Egypt. Nutr. Res. Dep., Natl. Res. Cent., Cairo, Egypt. Zeitschrift fuer Ernaehrungs wissenschaft (1976), 15(2): 224–30.

Hammouda, I.F.M. and Ahmed, Z.F., 1962. A phytochemical study of the seeds of *Pisum sativum.* I. "Drug Res. Unit, N. R. C., Cairo, Congr. Sci. Farm., Conf. Comun., 21, Pisa, Volume Date 1961, 551–564.

Hedges, L.J. and Lister, C.E., 2006. The nutritional attributes of legumes. Crop and Food Research Confidential Report No. 1745.

Hsieh, H.C., Kao, T.H. and Chen, B.H., 2005. Fast HPLC method for analysis of isoflavones. *Journal of Liquid Chromatography and Related Technologies,* 27(2): 315.

Jial, W., Gao, W. and Tanf, L., 2003. Antidiabetic herbal drugs officially approved in China. *Phytother. Res.,* 17: 1127.

Kundsen, I.M., 1986. HPLC determination of oligosaccharides in leguminous seeds. *J. Sci. Food Agric.,* 37: 560.

Lippiello L., Nardo, J.V., Robert Harlan, R. and Chiou, T., 2008. Metabolic Effects of Avocado/Soy Unsaponifiables on Articular Chondrocytes. *Evid Based Complement Alternat Med.* June, 5(2): 191–197.

Murai, A., Iwamura, K., Takada, M., Ogawa, K., Usui, T. and Okumura, J., 2002. Control of postprandial hyperglycemia by galactosyl maltobionolactone and its novel anti-amylase effect in mice. *Life Sci.,* 71: 1405–1415.

Obukowicz, M.G., Dean, J.W., William, J.S., Cynthia, L.M., Kevin, S.C., Kevin, L.D., Amiram, R., Philip, N., 1998. Novel, Selective Δ6 or Δ5 Fatty Acid Desaturase Inhibitors as Antiinflammatory Agents in Mice. *Journal of the Pharmacology and Experimental Therapeutics,* 287(1): 157.

Patricia, O., Deborah, P.D., Russell, W.C., John, G., Parastoo, A., Tony, B. and Moshe, B., 1998. Identification of a novel triterpenoid saponin from *Pisum sativum* as a specific inhibitor of the diguanylate cyclase of *Acetobacter xylinum. Plant and Cell Physiology*, 39(2): 144–152.

Perez, C., Paul, M. and Bazerque, P., 1990. Antibiotic assay by agar-well diffusion method. *Acta. Boil. Med. Exp.,* 15: 113–115.

Sabahat, S. and Perween, T., 2005. Antibacterial activities of *Mentha piperita, Pisum sativum* and *Momordica charantia. Pak. J. Bot.,* 37(4): 997–1001.

Shayne, C.G. and Christopher, P.C., 1992. *Animal Models in Toxicology*, 2nd Edn. Marcel Dekker Inc. Publisher, United Kingdom, p. 39.

Singh, K.N., Mittal, R.K. and Barthwal, K.C., 1976. *Ind. J. Med. Res.,* 64: 754.

Syam, M., Ahmad B., Abdul-Siddig, I.A., Al-Zubairi, A.S., Elhassan, M.M., Yousif, M., 2008. Antibacterial and antioxidant activities of *Typhonium flagelliforme* (Lodd.) Blume Tuber. *American Journal of Biochemistry and Biotechnology,* 4(4): 402–407.

Tiwari, P., Singh, A., Singh, U., Maurya, S. and Singh, M., 2009. Chromatographical analysis of phenolic acids in different preparations of pea (*Pisum sativum*) and chickpea (*Cicer arietinum*). *The Internet Journal of Alternative Medicine,* 8(1).

Troszynska, A. and Ciska E., 2002. Phenolic compounds of seed coats of white and coloured varieties of pea (*Pisum sativum* L.) and their total antioxidant activity. *Czech J. Food Sci.,* 20: 15–22.

Winter, C.A., Risley, E.A. and Nuss, G.W., 1962. Carrageenan-induced oedema in hind paws of the rat as an assay for anti-inflammatory drugs. *Proceedings of the Society for Experimental Biology and Medicine,* 3: 111, 544–546.

Zheng, C., Yoo, J., Lee, T., Cho, H., Kim, Y. and Kim, W., 2005. Fatty acid synthesis is a target for antibacterial activity of unsaturated fatty acids. *FEBS Letters,* 579: 5157–5162C.

Annex 1

7,4′-dihydroxy isoflavone (**daidzein**) (**1**): pale yellow prisms, Rf= 0.35 (EtOAc/toluene/formic acid; 50:50:10), mp 315-325°C. ^{1}HNMR (CD3OD-300MHZ): 10.7 (s, 1H, OH-7), 9.5 (s, 1H, OH-4′), 8.38 (s, 1H, H-2), 7.97 (d, *J*=8.1Hz, 1H, H-5), 7.43 (d, *J*=8Hz, 2H, H-2′,6′), 6.94 (dd, *J*=8.2Hz, 1H, H-6), 6.86 (d, *J*=2Hz, 1H, H-8), 6.81(d, *J*=8.4Hz, 2H, H-3′,5′).

Annex 2

4′,5,7-trihydroxy isoflavone (**genistein**) (**2**): white powder, Rf= 0.40 (EtOAc/toluene/formic acid; 50:50:10), mp 252-254°C. ^{1}HNMR (CD3OD-300MHZ): 10.7 (s, 1H, OH-7), 10.3 (s, 1H, OH-5), 9.5 (s, 1H, OH-4′), 8.31 (s, 1H, H-2), 7.35 (d, *J*=8Hz, 2H, H-2′,6′), 6.37 (d, *J*=2Hz, 1H, H-6), 6.26 (d, *J*=2Hz, 1H, H-8), 6.79(d, *J*=8.4Hz, 2H, H-3′,5′).

Annex 3

7-hydroxy-4′-methoxy isoflavone (**formonetin**) (**3**): yellowish white crystals, Rf= 0.65 (EtOAc/toluene/formic acid; 50:50:10), Rf= 0.20 (Toluene/$CHCl_3$/EtOH; 40:40:10), mp 220-225°C. ^{1}HNMR (CD3OD-300MHZ): 8.35 (s, 1H, H-2), 7.98 (d, *J*=9Hz, 1H, H-5), 6.94 (dd, *J*=8.2, 2.1Hz, 1H, H-6), 6.87 (d, *J*=2.1Hz, 1H, H-8), 7.51 (d, *J*=8.4Hz, 2H, H-2′,6′), 6.91 (d, *J*=8.5Hz, 2H, H-3′,5′), 3.79 (s, 3H, OCH_3-4′).

Chapter 2

Smart Advanced and Functional Materials and Structures: Design, Developments and Applications

*U.S. Tandon**

Planning and Performance Division,
Council of Scientific and Industrial Research,
2, Rafi Marg, New Delhi – 110 001, India

ABSTRACT

The creation and harnessing of advanced materials to develop a useful system is an upcoming stream of multidisciplinary research involving physics, chemistry, life sciences, materials science and electrical engineering. Some advanced materials like metamaterials are formulated through bulk processing or by using special processes such as plasmonics at nanometric dimensions. Some others, for instance those used in controlled delivery of chemicals or phonons are prepared through functionalization. Most of these techniques involve surface modification of materials made at nano-dimensions, where completely new characteristics manifest in the conventional materials. These functionalities can be embedded, coated or weaved in the conventional system. The fabrication of smart metamaterials is a tough challenge; nevertheless it opens immense avenues as was done by quantum mechanics.

Microcracks in conventional materials arising from stress generated cleavage of bonds or an external impact affect the material characteristics and lead to the failure of a system. By taking a cue from the biological systems, endowed with inherent ability to heal damages within their internal system or on the surface, a new group of advanced materials–Self-healing materials–is being developed. Composite, ceramic or metal is impregnated with smart ingredients to execute the self-healing. These materials detect a microcrack within its bulk and repair it to restore the original properties. The resulting extension of reliability and

* E-mail: tandonus@yahoo.com

efficiency at inaccessible locations like space, off-shore, artificial bones, bridges etc. can save the costs and facilitate missions to newer destinations. Applications of hyper-pure and ordered crystalline structures like silicon and diamond are well known. However, amorphous alloy made by rapid cooling of the melt have shown better toughness, corrosion resistance, electrical resistivity and the ability to prevent the loss of stored magnetic energy. Its use as the core of transformer delivers a saving of standby power requirement and reduces greenhouse gases.

Keywords: *Amorphous alloy, Functionalize, Magnetic energy, Metamaterial, Microcrack, Plasmonics, Self-heal, Smart material.*

Introduction

The advent of new materials such as fibres, polymers, semiconductors, ceramics etc. complimented technological forays in new disciplines like aerospace, IT, biomedicine, civil construction and their mass usage. Fusion of knowledge across disciplines has offered tools including the nanotechnology. It is imperative to harness them to realize advanced materials.

Innovative design and assembly of new materials was bestowed with a huge fillip with the downsizing of conventional materials to nanometric clusters. At these dimensions, their physical properties such as reactivity, conductivity and response to photons exhibited astonishing transformation. It gave a thrust to purposive research leading to new systems like Atomic Force Microscope and Scanning Tunneling Electron Microscope (AFM and STEM) which manipulate and monitor materials at those dimensions. While the design of system fetched Nobel Prize, intense and focused research resulted in the development of reliable yet least complicated techniques of creating nano-structures of metals and compounds (Tandon, 2003). The chemical, biomimetic, self-assembly and other avenues using for instance condensation, cellular biosynthesis, and adhesion at moderate or ambient pressure thus got established. Some of the advanced materials such as biominerals, composites in various matrices and core-shell structures finding applications in catalysis, communication and medicine have emanated from this route.

The composites of polymer and nanoceramics are being designed and synthesized into scaffolds for the growth of natural tissue such as bone and teeth. These nanostructures control interactions at cellular dimensions and pave a path to regenerate tissues via biomimetics. Nano-scaffolds are self-assembled structures of composites that mimic biomineralization and tend to be far more biocompatible than conventionally engineered structures, facilitating tissue growth around itself. Similarly, the composites made using nanoparticles of noble and transition metals placed in interlamellar spacing of porous clay (preferably bio-degradable) have shown potential applications in drug delivery, pollution control and antimicrobial systems.

The compliment of information science has provided intelligent abilities such as sensing, processing and actuation to the erstwhile dumb materials and thus converted them into advanced materials, which are often called Smart materials. Nucleating with tiny structures and scaling up to large systems, the functional properties of advanced materials endow them with either of the following abilities: a) to identify and amend in response to external ambient or b) to diagnose and remedy itself in response to its internal ambient. Tiny structures with the above functionalities can be integrated to create advanced materials. These are in turn harnessed into smart system that can judge the stimuli like stress, magnetic field, pH or UV and manifest a response such as harden retract or disperse and they often imitate a biological system. The laboratories of Council of Scientific and Industrial Research (CSIR), India are developing and functionalizing (Tandon, 2011) a variety of nanomaterials for specific applications.

Amorphous Alloys (Metallic Glasses)

Conventionally, alloys and metals are made by the solidification of its molten state through a process of gradual cooling. This process allows its atoms to arrange themselves in a crystallographic order through nucleation and growth, forming single crystal grains of different dimensions separated by boundaries, which jointly determine its physical characteristics. However, if the entire process of solidification is executed in a short span of time, say millisecond, the atoms do not get to reach their location decided by crystallographic order before the melt gets frozen. This amorphous or glassy state is often achieved by hastening the process of quenching its heat. It is done by reducing the volume of molten bulk to be frozen. Alternatively, atoms of different dimensions and chemical character can be added to the melt to reduce its freezing temperature and thus the alloy can be solidified without crystallization.

Such an amorphous alloy called metallic glass was fabricated at Caltech, USA when Duzez cooled Au-Si alloy so rapidly that it remained amorphous at room temperature. It has fewer grain boundaries than the conventional metals. So, metallic glass exhibits better corrosion resistance and it is not brittle. Conventional metals can be deformed permanently whereas the metallic glass regains its original shape far more easily than most other materials. Conventional glass is known to be strong, hard, and non-deformable. So, it cracks easily whereas metals do not crack easily but have limited strength. Metallic glass combines the strength of glass with the toughness of metals. The strength, elastic limit and toughness of metallic glass are two to four times that of steel or titanium. Just like plastics, the metallic glass softens on heating which facilitates its molding and opens new avenues for its applications. It is already being used commercially to make powerful golf club heads. It is shown in Figure Figure 2.1. Several other applications have emanated and some of these are: Aeroplane parts, Super-tough cases for cell-phones, dry erase whiteboards, and bioactive materials implantable in bones and as superconducting glass to contain plasma in fusion reactors. Crystallographic order limits the accuracy of finest structure made using conventional materials. Its absence in metallic glass renders them ideal for replicating surfaces and structures with dimensions like 10 nm used in nano molds and nano-imprinting. Metallic glasses have high shear modulus making them easy to deform but not crack. Their low Young's modulus and high yield stress render large resilience that makes them very good spring materials.

This alloy with abundance of inexpensive Titanium (Ti) is ductile (deforms without breaking) and tough (resists cracking) like a Zirconium (Zr) based alloy. Its density is 4.5–5.0 gm/cc whereas Zr based alloy has density 5.6–6.4 gm/cc (too heavy for aerospace).

Figure 2.1: Amorphous Alloy Sample

***Source*: Douglas Hofman, Caltech Proceedings of the National Academy of Sciences.**

Glasses including those of metals are prone to crystallization upon melting and losing their glassy properties. By mixing atoms of different sizes such as Ni or Zr at the stage of molten alloy, efforts are being made to raise the melting point and avoid crystallization. Fabrication of metallic glasses that can withstand high temperature is desired for civil applications such as in engines and in strategic applications like missiles made to encounter metallic objects including armours. Missiles made from conventional metal or alloy is likely to flatten upon impact at high temperature but a container of metallic glass with high melting point is likely to sharpen (like other glass) and enhance its piercing ability. However, the stringent requirement of rapid cooling (10^6 °C/sec) to make metallic glass from molten alloy stage has been limiting its fabrication in the bulk form so far. But metallic glass in the form of powder, wire or long thin ribbons have been successfully made. One of the frequently used techniques to prepare ribbons of amorphous or glassy state is melt spinning shown in Figure 2.2a (Narsimhan 1979). Here, the alloy in its molten state is poured on a fast spinning metallic-drum where it spreads and gets frozen by planar flow casting to deliver continuous ribbon with a controlled thickness of about 30mm. Figure 2.2b shows the metallic glass ribbons prepared at CSIR-NML. The thickness and width of ribbon could be controlled (Sarkar *et al.*, 2010) by the parameters of melt spinning process which alleviates rolling. The desirable properties of the metallic glass can be tailored by iterating the composition of the alloy.

The obvious way of making components from metallic glass is to melt the alloy, cast it in a metal mold and cool it quickly before it crystallizes. But it fell short of practicality for two reasons: a) lack of techniques for heating/cooling the melt so quickly, and b) the molds are not likely to withstand such high temperatures. But it has been found that metallic glasses become moldable by acquiring the requisite fluidity at a temperature that is half of its melting point. Recently, Johnson *et al.*, 2011 of Caltech developed rapid discharge forming technique utilizing Ohmic heating by applying intense electric pulse delivering a Mwe of power in 1 msec. They could heat

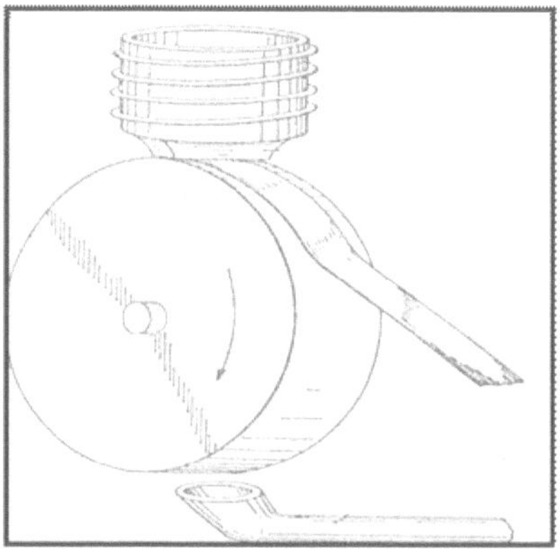

2.2a

2.2a

Figures 2.2a and 2.2b: Metal Spinning System and Thin Ribbon of Amorphous Metal

a metallic glass rod of dia 4 mm and length 20 mm at a rate of a 10^{6}°C/sec raising its temperature to 550 °C in 500 microseconds. By injecting this softened glass in a donut-shaped mold and freezing it before it crystallizes, flow defects and oxidation were avoided. Precisely shaped parts of metallic glass can thus be formed using a technique, akin to plastic forming and ensuring that it is still amorphous and its characteristics as metallic glass are intact. Schroers *et al.*, 2006 prepared a micro gear with a cog width of about 5 µm. The accuracy in replicating component dimensions is ~10 nm, which is better than that for metals where the presence of microstructure is a limitation. Metallic glasses under tension at room temperature tend to become brittle due to plastic flow confined within the shear band. Composite materials have been developed by using metallic glass as matrix and introducing ductile fibre or ceramic whose size is comparable to the shear band and obstructs its propagation. Metallic glass composites have better strength, toughness etc. and efforts are showing way to cast and produce near net shape components of these composites.

Iron-based metallic glasses get easily magnetized by moderate magnetic fields and the lack of crystalline structure causes its electrical resistance to be higher than that of crystalline magnets. Their advanced ferromagnetic properties *viz.* the absence of magneto crystalline anisotropy and high electrical resistivity reduce their core loss and the eddy current loss respectively. Combined with improved corrosion resistance, this advanced amorphous material is ideal for making transformer cores esp. in hostile environment. For transformer applications, magnetic glasses made by mixing ~80 at per cent of a combination of transition metal like Fe, Co, Ni with small-atom metalloids, having chemical property between metals and non-metals, have been found most suitable. Thus ~20 at per cent of B, C, Si, P or Al are used in the alloy.

Metallic glass gets used in distribution and power transformers, motors, high frequency inductors, current transformers, and other devices requiring high permeability and low loss at low frequencies (<0.29 W/kg at 60 Hz, 1.35 Tesla). By replacing usual grain-oriented electrical steel with amorphous alloy as the core of a transformer, its "no-load loss" (or standby power requirement) drops by about 67 per cent to 80 per cent. Thus it helps reduce the emission of greenhouse gas (GHG). The switchover to transformers with metallic glass cores was made in USA in the 1980s and in Japan in the 1990s. China and India, groping for ways of sustainable development, are keen to use amorphous alloys as core of transformers for environmental compatibility as well as to enhance electric power efficiency. It is understood that the demand from these countries has grown to exceed that of Japan.

Manufacture of metallic glass moved from Allied signal to Honeywell to Hitachi Metals, with a production capacity of 50,000 Tonne/yr, partly in Japan and partly in USA. Vacuum Schmelze, Germany, AT and M Intl Trading Co, China, Samsung, TAG Heuer, Vertu Ltd., HEAD NV, Rawlings (Sport goods) are other producers of metallic glass. Strips from Metglas (Hitachi Metals in USA) are used as transformer iron core with extremely low core loss compared with grain oriented electrical steel. Defense Advanced Research Projects Agency (DARPA), USA. DARPA had a multi-billion dollar program on Amorphous Metals for: hull material for ships; corrosion resistant coatings; lightweight alloys for aircraft and rocket propulsion; wear-resistant combat vehicles. The government of Japan has also been funding research on metallic glasses for more than 30 years.

Owing to its advanced ferromagnetic properties and high electrical resistivity, the metallic glass in addition to being the ideal material for the core of distribution transformer, is finding application as quick-response magnetic sensors, audio/video magnetic tape heads, core of common mode chokes, security systems, SMPS, Telecommunications, Interface and Pulse Transformers etc.

Chromogenic Materials

Materials endowed with differential in the transmission or reflection of electromagnetic signal em across its spectrum; belong to the family of energy saving advanced materials. CSIR-CGCRI and Tata Chemicals Ltd India jointly introduced Gold-nano-triangle in SiO_2 to develop passive hard coatings (De *et al.*, 2007). The transparency of these material films increases by 50 per cent as the incident wavelength advances from infrared (1100 nm) to visible (600 nm). These results obtained by CSIR-CGCRI are shown in Figures 2.3a and 2.3b. Conversely, by utilizing nanoparticles that absorb infrared wavelengths as high refractive index materials energy saving coatings applicable to glass is also being developed. These coating reflect 20 per cent of incident radiation in visible range (600 nm) but 32 per cent of infrared (1000 nm).

This is a family of new materials that selectively control the spectral aspect of radiation in a switchable way. It happens in response to stimuli such as thermal, electrical, optical or pressure, pH ambient. Chromogenic smart materials have often been used as low-emittance coatings on glass sheets to lower the total heat flow through windows.

Such coatings, often of nanomaterials, are transparent to visible light but opaque to infrared and find use in transport systems including aircrafts, architectures, mirrors, and in large variety of information displays. Pyrolytic coating of transparent electrical

Figure 2.3a: Coated Glass

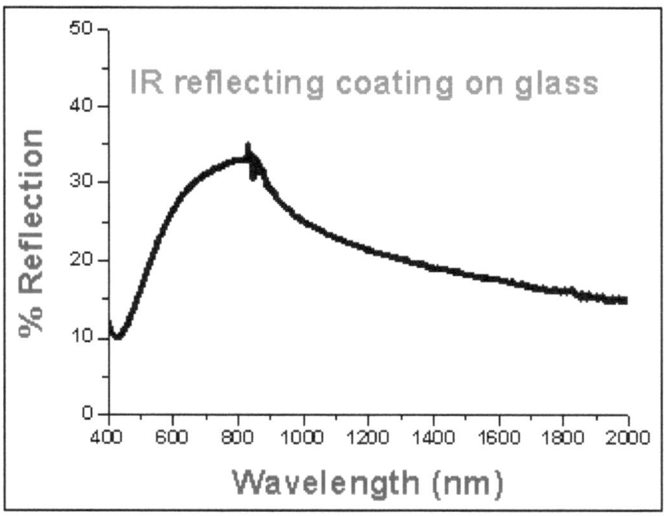

Figure 2.3b: Reflection Spectrum of Coated Glass

conductors like Indium doped tin oxide (ITO) and fluorine doped tin oxide have been used to suppress infrared transfer and impart thermal insulation. New organic semiconductors are an upcoming class of materials. Chromogenic materials include: a) electro-chromic materials that undergo reversible change in optical properties upon injection of light ions (an example is Liquid crystal display); b) thermo-chromic materials whose colour changes with ambient temperature, and c) photo-chromic materials, whose transparency changes with the intensity of ultraviolet (UV) illumination. The last has been used in ophthalmic products for many years.

Electro chromic materials are normally transparent but attain a dark blue or black or nearly opaque tint in response to a low electric voltage and remain so as long as the voltage is maintained. These could be films of MoO_3 or WO_3 or Mo/WO_3 deposited by CVD at low temperature and atmospheric pressure. Its coating on windows, upon darkening, causes absorption and radiation of incident sunlight/heat just like low emissivity surfaces. The requirement of continuous electric power to keep the window opaque has not deterred its use as the total power needed is generally small (~1.1 W/m^2 of window area) and in hot climates this consumption is offset by savings on endurant air conditioning (Figure 2.4). Alleviation of curtains and blinds is an added benefit since the privacy can be had by switching off the electricity. Long term savings on running cost by as much as 20 per cent is popularizing electro chromic insulating glass materials. Windows produced by Sage Electrochromics and Research Frontiers have already found use in architecture and high end cars respectively.

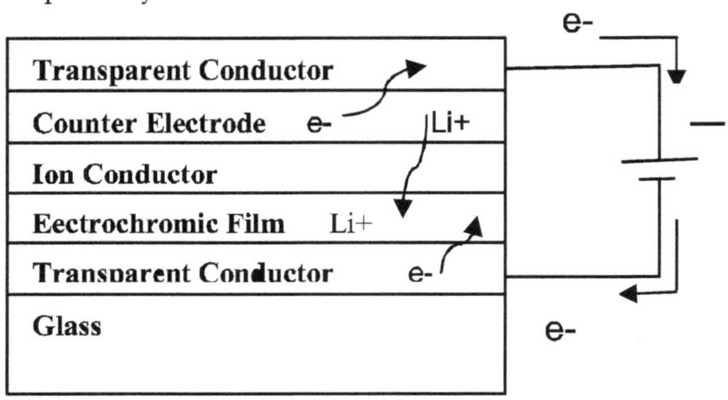

Figure 2.4: Functioning of a Electrochromic Window [Sage Electrochromics]

The antiglare mirrors in high end cars using Electrochromics providing dynamic switching have an estimated market of $M300/year. Companies such as Gentex, Magna Donnelly, Nikon, and Murakami-Kaimedo share this market. Saint-Gobain is commercializing switchable glazing for the roof (Giron, *et al.*, 2003); its estimated price for one m^2 may be above $100. It is commercializing electro chromic window (Betelle *et al.*, 1999) that brings down the glare through windows and visor in an aircraft, reducing the contrast as well as transmitted light to a few percent of the ambient. Chromogenic visors that switch at 1.6V for helmets worn by two-wheeler riders have been developed by Upsaala University. Alkali hydride is also being investigated as switching chromogenic material. The optical density of a nanocell p-n junction made using dye-sensitized anatase TiO_2 can be controlled through its junction voltage. Such nanocells are being developed as electro chromic devices supported by photovoltaic cells.

Thermo chromic materials often based on transition metal oxides like VO_2 change their color when exposed to heat. Transparent substrates with a coating of passive thermo chromic material exhibit a transition in their heat reflectivity with changing temperature and become opaque at high temperature. Such materials find critical use in aerospace industry to enhance the emissivity of heated surfaces. Transition from metal to insulator for $V_{1-x}Nb_x0_2$ can be tuned between 64°C and -4°C by tailoring the value of x. Thus, a window that absorbs heat in cold ambient starts reflecting heat when the ambient turns hot. This transition takes longer time than the switching by an electro chromic window but the former requires no electricity. However, additional resistive heating of its surface can provide electrical control and faster response. In winters it lets IR come in and assists indoor heating whereas in summers it reflects them away, curtailing the indoor heating. They are often made using hydro gels or polymer combinations such as polyether/ethylene oxide/carboxyvinyl. Some organic thermo chromic inks are getting used as temperature indicators.

An innovative use of thermo chromic materials is Heat Mirror, which can be formed as follows. A coating of TiO_2-Ag-TiO_2 typically lets in wavelengths between 300 and 650 nm and reflects longer wavelength especially the far IR. If a layer of heat absorber is applied on the inner surface of the window, it will hold the heat and restrict its re-radiation in the form of IR with longer wavelengths (Figure 2.5). A combination of tunable $V_{1-x}Nb_xO_2$ as the outer pane and TiO_2-Ag-TiO_2 as the inner pane produces smart window reflecting solar energy in summer but transmitting solar energy in winter by preventing re-radiation of long wavelength heat.

Self Healing Materials

Materials used to construct load bearing and shape-holding structures are required to be tough and strong. However, inherent stress and external stimulus like

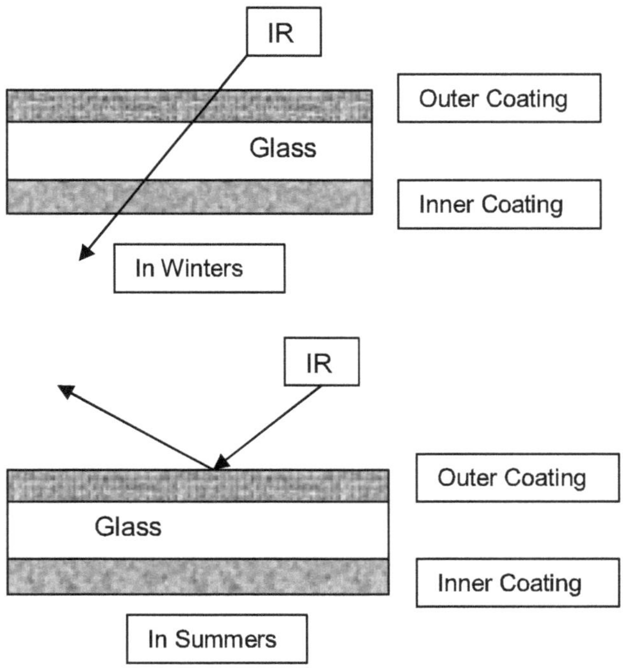

Figure 2.5: Heat Mirror made Using Coating of Thermo Chromic Materials

those arising from usage or impact, gradually wears out most materials and they undergo microscopic damages. Generally, it results in the cleavage of bonds, whose propagation manifests in micro cracks that keep on expanding and eventually results in catastrophic fractures causing a complete failure of the structure. It eventually limits the life-span for performance to full specifications of critical man-made structures like bridge or satellite and prevents mission to destinations that require longer endurance.

Taking a cue from living organisms and other biological systems that are endowed with inherent ability to heal damages within their internal system or on the surface, a new group of advanced materials–Self-healing materials–is being developed. Fruitful amalgamation of diverse disciplines like physics, chemistry, polymer science, nanotechnology, environmental engineering etc. has facilitated (White *et al.*, 2001) this novel imitation of biological capability to materials. At the two-dimensional level, self-healing paint that repairs scratches and inhibits corrosion of the surface underneath has been developed by Nissan. Arkema, France offers self-healing layer that resists scratches on consumer products. Spurt in industrial products based on self healing materials is imminent in view of their demonstrated potential. Monomers stuffed in a three-dimensional material could repair its cracks by polymerizing through external stimuli such as heat.

To accord the self-healing ability, we may introduce meso-capsule filled with healing Agent like dicyclopentadiene and suitable catalyst in the body of the material. The energy released by a spreading micro-crack ruptures few capsules, releases the healing agent that contacts the embedded catalyst (Figure 2.6) such as Grubbs'. Resultant cross linking reaction among pendant groups bonds the crack restoring up

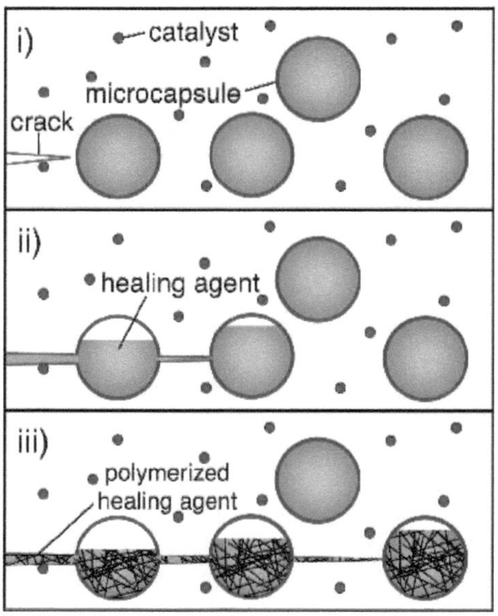

Figure 2.6: Crack Ruptures Microcapsules Releasing Healing Agent that is Catalyzed to Bond

to 75 per cent of the strength of the original composite, for instance a tyre. In another avenue, hollow fibers, hundred times thinner than the capsule–some filled with monomer and some with hardener work the same route, avoiding the introduction of unused catalyst. Thermoplastics damaged by collision can be self healed using the heat of collision and retro Diels-Alder reaction reforming diene, alkenes and favouring rearomatization.

Plastics, paints, metals, alloys, ceramics and composites can all be made self-healing. Recent demonstrations of executing self healing operation while the structure is in operation have shown that up to 96 per cent of its original strength as well as acoustical, thermal, electrical properties can be regained. It also enhances the toughness several-folds. The in-built circulation system of self healing materials can be used for thermal exchange, opening up further applications. Fibre-optic network can be embedded to detect damage and delivering stimulus for healing.

Spurt in industrial products based on self healing materials is imminent in view of their demonstrated potential for such a widespread application. The release of initial critical applications in defence/aerospace may be deferred until the technology matures and gets harnessed in civil applications too. Almost 20 million tons/year of composite material is used in defence, engineering, electronics, biomedicine and remote locations. Self-healing technology will find applications nearly everywhere–in space, aviation, transport, civil engineering, electronics, artificial bones, dental implants, roads, and so forth.

Electro- and Magneto-Rheostatic Materials

Some fluids in oily stage undergo a dramatic change in their rheological properties in response to an electric or magnetic field and get transformed to highly viscous or nearly-solid state in a few milliseconds. This transformation of state gets completely reversed when the applied field is withdrawn. The former materials are termed Electro-rheostatic (ER) and the later as magneto-rheostatic (MR) materials. While oils, cornstarch and milk chocolate exhibit ER properties, suspension of Iron particles of size 3-5 micron in carrier liquid such as oil make it magneto rheostatic. ER fluids are ideal for noise muffling or reducing quivers in engine, clutch, brake and also in bulletproof vests. MR fluids effectively diminish severe vibrations like those occurring in grinding-polishing machine, aerobic systems, car suspensions, from buildings in tremor zone and even prosthetic attachments. Iron microparticles suspended in MR fluids are likely to settle through gravitation and reduce its rheological response to magnetic field. It is prevented by adding surfactants or stabilizers that enhance lubricity. MR fluids solidified by very strong magnetic field have strength like 100 kPa. They are an order stronger than ER materials and do not get easily spoilt by contamination.

Metamaterials

Materials have some fundamental characteristics like mass and dimensions. They also have intrinsic electromagnetic parameters like permittivity and permeability. Permittivity (ε) is a measure of electric polarizability or the ability of the material to permit an applied electric field throug it. Permeability (μ) is a measure of the degree of

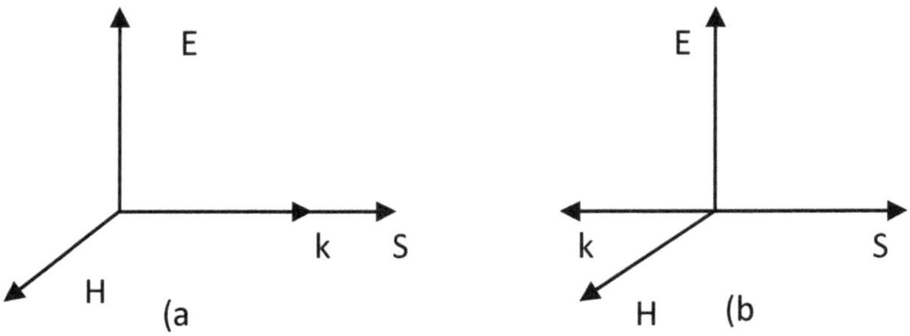

Figure 2.7: Electric, Magnetic, Poynting and Propagation Vectors for Normal and Metamaterials

magnetization that a material undergoes in response to an applied magnetic field. The square root of the product of ε and μ, after their normalization to the permittivity and permeability of free space respectively, is the refractive index of the material. The value of ε and μ for naturally occurring materials are positive. Electromagnetic metamaterials however – by definition – have negative values for both ε and μ. It makes their refractive index negative (Shelby *et al.,* 2001) and reverses the role played by a convex/concave lens.

When electromagnetic wave propagates, its electric (E), magnetic (H) and propagation (k) vectors are at right angles to each other and for most materials the Poynting vector (S) representing the flow of em energy is parallel to S as in Figure 2.7(a). However for metamaterials the Poynting vector S is antiparallel to the propagation vector k as in Figure 2.7(b). When the energy and propagation are in opposite directions, the Doppler effect and Cherenkov radiation get reversed too (Jha *et al.,* 2007 of CSIR-NAL). The realization of such an unconventional material eluding established wisdom remained hypothetical for some years although Pendry *et al.,* in 1996 predicted plasma oscillations at low IR and GHz frequencies.

The classification of materials based upon their electromagnetic properties is shown in Figure 2.8. When em wave enters a material, its electric and magnetic fields exchange energy with the electrons in the material, which in turn manipulates the propagation of em wave.

The plasma frequency of a material at which its free electron density oscillates is often in the ultraviolet range and all incident em signals of smaller energy are reflected by the material. Additionally, since the wavelength of em signal of our interest (UV, visible, IR and microwave) is far larger than the atomic dimensions, most materials are 'seen' by the incident signal as a continuum. Incidentally, the interaction properties *viz.* refraction or reflection of electromagnetic waves from a material can be modified, though slightly, by varying its plasma frequency by chemical modulation of its molecular composition. But a more feasible and effective way has emerged through the techniques of nanofabrication, which facilitate larger transformation in mechanical properties than by chemical synthesis.

If we assemble a structure, the dimensions of whose constituents and their spacing are comparable to the wavelength of incoming electromagnetic signal, the

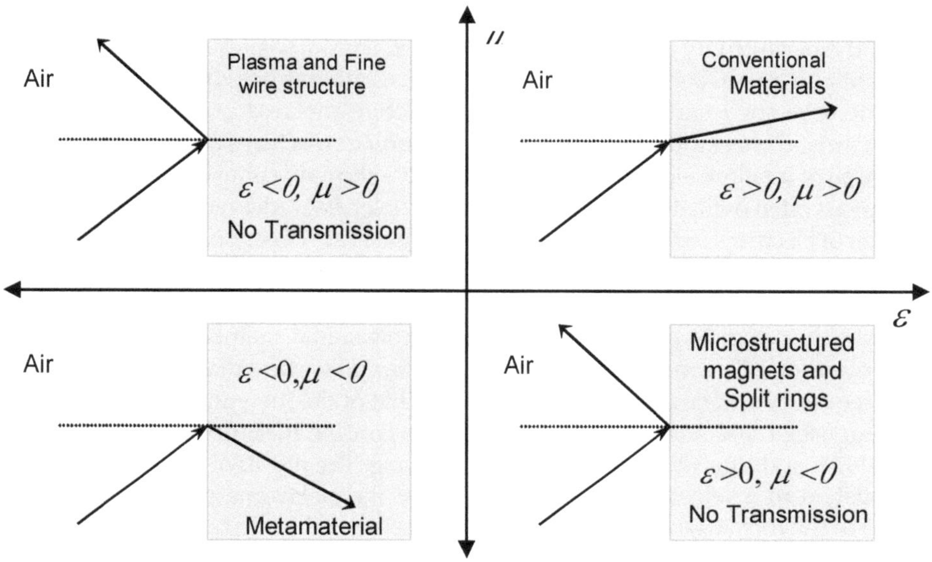

Figure 2.8: Classification of Materials Based on Material Parameters

latter will fail to resolve the constituents of this structure. Thus, the incident em signal will interact with this structure like a new actual material. Such a structured material having electromagnetic properties far different from those of the natural materials is called metamaterial. These are photonic composites fabricated using natural material such as Gold, Copper and insulators in specifically designed shapes and patterns which ensure that their electromagnetic properties are decided not only by its molecules but also by the size and shape of the assembly. We will see how this metamaterial can be built by arranging the geometry of the atoms and molecules at sub-wavelength dimensions so that the composite delivers negative magnetic permeability and optical magnetism.

Zero reflection desired for invisibility cloaking is often interpreted as absolute transmission (or absorption), which is nearly impossible to achieve. The metamaterial structure, which is smaller than the wavelength of the em signal, guides the latter around itself just (like the flow of water around a surface) and back to its original path. It prevents the formation of any shadow and also the detection of the structure. Cloaking metamaterial structures of sub-wavelength size can be made for microwaves, whose wavelength at 100 Ghz is 3mm. However, wavelength of visible spectra is four orders smaller (400-700 nm) and structures much finer than that can only be made and characterized using nanolithographic tools.

Two dimensional perforated stacks of selected metal-insulator combination (Zhang *et al.*, 2006), thicker than the incident wavelength have shown metamaterial properties. Some of the polymer based nanocomposites have exhibited the properties of metamaterials. For instance, core-shell Fe/Fe_3O_4 magnetic nano particles or those formed by incorporating nanodimensional carbon fillers in the poly-ether-imide matrices. Its dielectric properties and transmission-line perturbation would confirm whether the permittivity and the permeability of such composites are negative. CNT coated with Ni, metal coated carbon nanotube in combination with normal CNT or and some core-shell nanomaterials are prospective metamaterials. However, to fabricate three-dimensional metamaterials, plasmonic structures far thinner than the wavelength of incident electromagnetic wave have shown promising success. While the plasmons often oscillate at a frequency in ultraviolet and reflect smaller frequencies, the packet of electrons induced by em signal may generate Plasmons that are confined to a metallic surface and oscillate at a frequency in the visible region or IR region. By designing nanodimensional structure of optimized dimensions using thin films like Gold and Silicon nitride, photons in the incoming em signal are made to interact with surface plasmons. The incident photons thus generate surface plasma polaritons of higher frequency and smaller wavelength than that of the incoming signal. As their name suggests, these surface plasma polaritons confine their flow within the thin metallic layer and thus the em signal is guided along. The negative permeability thus accomplished in a selected frequency region of the electromagnetic spectrum is sensitive to the formation of Surface plasma polariton, which in turn is governed by the dimensions of the nanostructure.

An illustration of such plasmonic material with negative magnetic permeability and optical magnetism is photonic composites in the form of Split Ring Resonator (SRR) whose multilayer nanodimensional structure was fabricated (Na Liu *et al.*, 2008) by making use of nanolithgraphic techniques (Figure 2.9).

A typical split resonant resonator has alternate layers of Gold ~200 Å thick sandwiched in somewhat thicker layer of non conductor in the shape of alphabet C in a rectangular configuration. The two parallel arms of the structure are ~3800 Å and the arm joining them is ~4200 Å. The width and thickness of the pattern are ~800 Å and ~200 Å respectively. Blanket thin films of Gold are deposited by evaporation

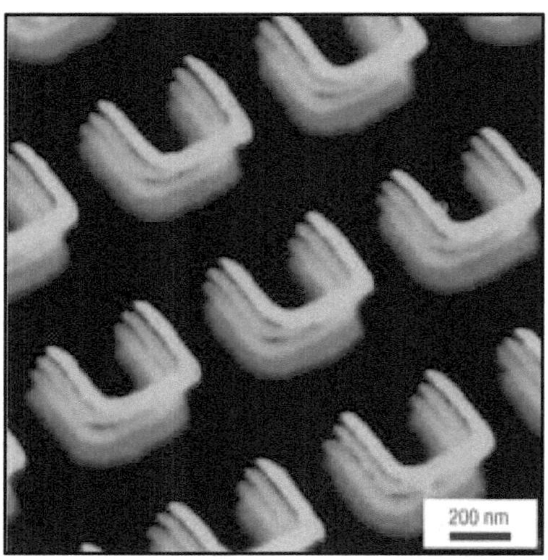

Figure 2.9: Plasmonic Metamaterial Structure Under Fabrication (Na Liu *et al.*)

whereas non-conductors such as polymer can be spun onto it. Various unit processes like mask making, lithography, planarization and etching are required to fabricate this structure. The permittivity is found to undergo a distinct dispersive behavior in the range 5000–3750 nm and its real part drops from strong +ve to strong –ve value. Thus, it is possible to build a material with desired optical properties (Shalave 2007) and also to control them.

Electromagnetic metamaterials with –ve refractive index can be used as "Superlens" for visible-light imaging of molecular and atomic scale objects. Lenses of conventional materials with +ve refractive index, owing to their diffraction limit, cannot resolve objects small than ($\lambda/2$) of visible light. However, plasmonic materials reduce the wavelength of incident light to a fraction of its original. Thus, a superlens picks up spatial information in regions smaller than the optical wavelength. The unique properties of metamaterials at THz, IR and visible wavelengths are being tapped for invisibility cloaking (Schurig *et al.*, 2006) and superlensing (Wiltshire *et al.*, 2006). Light falling on conventional materials, with a positive index of refraction, exerts a positive pressure, meaning that it can push an object away from the light source. In contrast, illuminating negative index metamaterials should generate a negative pressure that pulls an object toward light. Plasmonic metamaterials can also be used to make optical switches, photodetectors, modulators, directional light emitters and high sensitivity biological, chemical and optical sensors by exploiting super-confinement of surface plasmons. Specific functionalities could be embedded, coated or weaved into system made of conventional materials such as cloth, wall or vehicle making them sense, respond intelligently and remain lightweight. The fabrication of smart metamaterials with above properties like negative refraction continues to be a tough challenge; nevertheless it is opening immense avenues as was done by quantum mechanics.

Acknowledgements

I am grateful to Prof. S.K. Brahmachari, DG CSIR and Dr. Sudeep Kumar, Head PPD for their precious encouragements. Discussion made and inputs obtained from Dr. G. De, CSIR-CGCRI, Dr R.M. Jha, CSIR-NAL and Dr. A. Mitra, CSIR-NCL are gratefully acknowledged. I am also grateful to all the scientists in different research laboratories of CSIR whose findings have been integrated.

References

Betelle, F. *et al.*, 1999. Highly durable all-solid state electrochromic glazings. In: *Switchable Materials and Flat Panel Displays*, Proceedings of SPIE, Lampert, C. M. (eds.) 70, pp 3788.

De, G., Sastry, M. and ShivShankar, S., 2007. Coating to reduce near infrared radiation, method of manufacturing and applying coating. Indian Patent Application # IN200700747.

Duwez, Pol, Willens, R.H. and Klement, W., Jr., 1960 Continuous series of metastable solid solutions in silver-copper alloys. *Journal of Applied Physics*, 31(6): 1136–1137.

Giron, J.-C., *et al.*, 2003. Electrochromic automotive sunroofs. *Proceedings of Glass Processing Days*, pp. 460.

Jha, R.M. and Wiesbeck, W., 2007. Trends in metamaterials evolution. Validation and Applications Part I and II, Technical Report, Universitat Karlsruhe.

Johnson, W.L. *et al.*, 2011 Beating crystallization in glass-forming metals by millisecond heating and processing. *Science*, 332(6031): 828–833.

Schroers, J., Nguyen, T. and Desai, A., 2006. IEEE. *MEMS*, 26: 298.

Na Liu, Hongcang, G., Liwei, F., Kaiser, S., Schweizer, H. and Giessen, H., 2008. *Nature Materials*, 7: 31–37.

Narasimhan, M.C., 1979 Allied signal made the metallic glass ribbons commercially viable by producing continuous ribbons by planar flow casting US Pat: 4,142,571.

Pendry, J.B., Holden, A.J., Stewart, W.J. and Youngs, I., 1996. Extremely low frequency plasmons in metallic mesostructures. *Phys. Rev. Letters*, 76(25): 4773–4776.

Sarkar, P., Mohanta, O., Pal, S.K., Panda, A.K. and Mitra, A., 2010. Magneto-impedance behavior of Co-Fe-Nb-Si-B-based ribbons. *J. Magn. Magn. Materials*, 322: 1026.

Schurig, D. *et al.*, 2006. Metamaterial electromagnetic cloak at microwave frequencies. *Science*, 314: 977–980.

Shalave, V.M., 2007. *Nature Photonics*, 1: 41–48.

Shelby, R.A., Smith, D.R. and Schultz, S., 2001. Experimental verification of a negative index of refraction. *Science*, 292: 77–79.

Tandon, U.S., 2003. Strides in nanotechnology by India. *International Congress and Exhibition on Nanotechnology*, Nanotech2003+Future, Japan.

Tandon, U.S., 2011. Nano-structures, nanomaterials and systems. *Nanotechnology: Present Status and Future Prospects in Developing Countries*, Published by NAM S&T Centre, NewDelhi, pp. 3–17.

Wiltshire, M.C.K., Pendry, J.B. and Hajnal, J.V. 2006. Sub-wavelength imaging at radio frequency. *J. Phys. Condensed Matter*, 18: L315–L321.

Zhang, S. *et al.*, 2006. Optical negative-index bulk metamaterials consisting of 2D perforated metal-dielectric stacks. *Opt. Express*, 14: 6778–6787.

White, S.R., *et al.*, 2001. Autonomic healing of polymer composites. *Nature*, 409: 794–797.

Abbreviations

CSIR–CGCRI: Central Glass and Ceramic Research Institute, Kolkata

CSIR–NAL: National Aerospace Laboratory, Bengaluru

CSIR–NCL: National Chemical Laboratory, Pune

CSIR–NML: National Metallurgical Laboratory, Jamshedpur

Chapter 3
A Review of Thermal Conductivity Properties of Carbon Nanocone to be Used in Re-Entry Vehicle

Morteza Rasouli Gandomani and Hamid Rashidi***

No. 53, 1st East Alley, North Haffez Street, Shahin Shahr, Esfahan, Iran

ABSTRACT

Since the discovery of carbon nanotubes (CNTs), the research community has paid much attention to this form of carbon nanostructure because of its exceptional properties. There are several methods for studying CNTs, such as molecular dynamics (MD), quantum mechanics and continuum theory, all of which have been variously used to determine the properties of CNTs, such as their mechanical properties, thermal properties, and electrical properties. However, another newly synthesized form of carbon nanostructure–the Carbon Nanocone (CNC)–has attracted less attention. Up to now, researchers have struggled to establish universally accepted approaches for CNC structures and growth mechanisms, and reports on their mechanical properties are even scarce. In recent years, by using optimized methods, about 20 percent of carbon nanocones, 70 per cent flat carbon plates and 10 per cent black carbon are obtained. This paper presents a survey into thermal conductivity properties of carbon nanocone materials to be used in re-entry vehicle by using molecular dynamics simulation. Today in aerospace fields, thermal conductivity of structures for re-entry vehicle is an important subject and in this paper reports on the introduction of carbon nanocone materials to be used in these structures. Also, this paper reports a survey on thermal conductivity properties of carbon nanocone materials to be used in re-entry vehicle by using molecular dynamics simulation. Today, carbon nanocone due to its unique properties has many applications in various industries including aerospace industries. One of the interesting and unique properties of carbon nanocone is lower conductivity properties of

E-mail: * rasooli@ae.sharif.ir; **hamidrashidi.1365@gmail.com these materials than carbon nanotubes in higher temperatures. Presented survey concludes that conductivity properties of carbon nanocone are based on their orientation. By molecular dynamics simulations, it is found that conductivity thermal properties of these advanced materials have nothing to do with the geometric parameter of length. Also, if initial temperature in re-entry phase increases, thermal conductivity will be reduced for these advanced materials. So, for the structures of re-entry vehicles, that must tolerate high initial temperatures, application of these advanced materials is very vital because in such temperatures, thermal conductivity property of these materials is very low and as insulation materials, they can be used in the structures of re-entry vehicles.

Keywords: *Carbon nanocone, Re-entry vehicle, Thermal conductivity, Graphite, Molecular dynamics simulation, Curvature, Graphene, Thermal properties, Fullerene, Rectification.*

Introduction

Carbon is unique among the elements for its ability that can be assumed as an extended two dimensional sheet structures in elemental form. Since the in-sheet bonding is extremely strong, these sheets are stable both as isolated objects (graphene), and when curved into cylindrical geometries (nanotubes) or quasi spherical geometries (fullerenes) wherein, *e.g.*, pentagonal rings provide the needed Gaussian curvature that is required points to form the curvature. An intermediate state between a sheet and a fullerene also exists, the carbon nanocone, wherein a single pentagonal ring or assembly of nearby pentagonal rings defines a conical apex, which is then extended by a pure-hexagon graphenic network into a larger conical structure. As it does for the nanotubes, fullerenes and graphene, the geometry of the nanocone defines a unique set of structural, mechanical, chemical, and electronic properties. One major

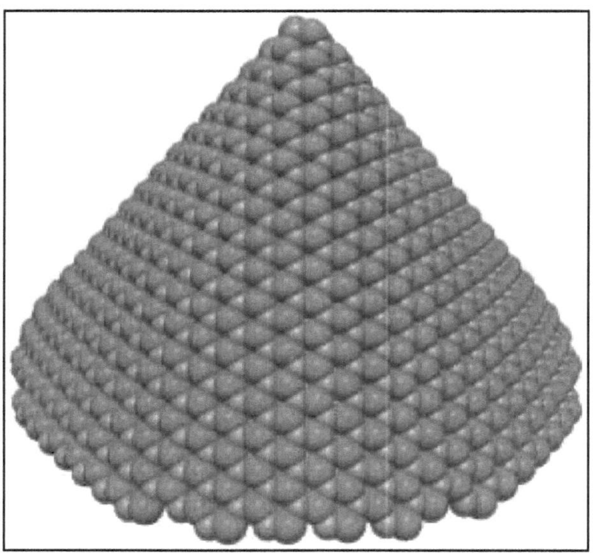

Figure 3.1: Carbon Nanocone with Apex Angle of 120 and Height of 30 Angstroms (Wei Jing Xi, 2007)

subclass of nanocones, nanohorns formed by ablating graphite with a CO_2 (Carbon dioxide) laser, are distinguished by high yield, high purity and an aggregate microstructure, features that have facilitated the development of potential applications in the adsorption and release of small molecules. In Figure 3.1 illustrates the one shape of carbon nanocone with apex angle of 120 and height of 30 Angstroms.

Carbon nanocones are produced in an industrial process that decomposes hydrocarbons into carbon and hydrogen with a plasma torch having a plasma temperature above 2000 °C. This method is often referred to as Kvaerner Carbon Black and Hydrogen Process (CBH) and it is relatively "emission-free", *i.e.*, produces rather small amount of air pollutants. At certain, well optimized and patented conditions the solid carbon output consists of approximately 20 per cent carbon nanocones, 70 per cent flat carbon discs and 10 per cent carbon black (Wei Jing Xi, 2007).

Plasma-assisted decomposition of hydrocarbons has long been known and applied, for example, for production of carbon fullerenes. Even if not optimized, it yields small amounts of carbon nanocones, which had been directly observed with an electron microscope already in 1994, and their atomic structure was theoretically modeled the same year (Wei Jing Xi, 2007).

The growth of carbon nanocone arrays on metal catalyst particles by deposition from a low-temperature plasma is studied by multi scale Monte Carlo/surface diffusion numerical simulation. It is demonstrated that the variation in the degree of ionization of the carbon provides an effective control of the growth kinetics of the carbon nanocones, and leads to the formation of more uniform arrays of nanostructures. In the case of zero degree of ionization (neutral gas process), a width of the distribution of nanocone heights reaches 360 nm with the nanocone mean height of 150 nm. When the carbon flux of 75 per cent ionization is used, the width of the distribution of nanocone heights decreases to 100 nm, *i.e.*, by definition a factor of higher degree of ionization leads to a better uniformity of the metal catalyst saturation and the nanocone growth, thus contributing to the formation of more height-uniform arrays of carbon nanostructures (Masako Yudasaka, 2008).

Cones can also be formed in vapor-phase carbon deposition onto graphite surfaces or in the pyrolysis of carbon-containing precursors such as hydrocarbons or CO (Carbon monoxide). Glassy carbons or naturally occurring graphite aggregates sometimes contain thicker conical structures. Such samples often contain heterogeneous mixtures of multilayered structures with nested cones. The end of a standard carbon nanotube also often forms a tapering structure, wherein a single pentagonal ring in a straight tube wall induces an inward taper, the tube eventually closing in a five-pentagon conical cap, with an opening angle of 20 Angstroms. An overall cone shape can also form in situations where it is not defined topologically by the presence of pentagons in the apex. Carbon cones can be classified at the coarsest level by the number of layers in the cone and the number of pentagonal disclinations defining the tip. Figure 3.2 illustrates schematically the construction of one-pentagon and two-pentagon cones in terms of wedges removed from a graphene sheet. More pentagons imply a sharper opening angle, in the sequence of approximately 113 per

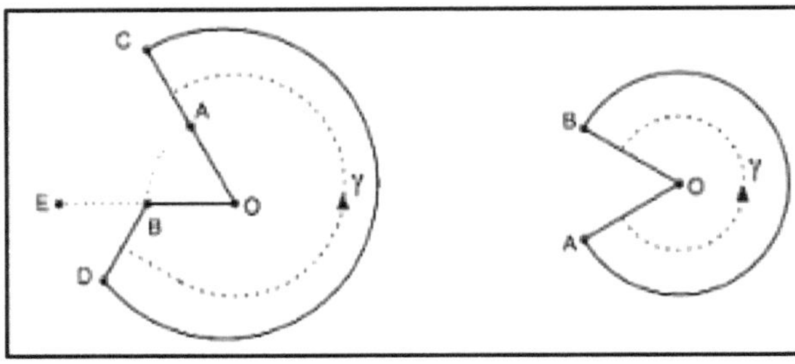

Figure 3.2: The Geometrical Construction of a Nanocone with One (*Right*) and Two (*Left*) Pentagons, in a Continuum Picture.

One or two wedges are removed from a full sheet to form the opening B-O-A or C-A-O-B-D. This opening is then stitched back together by joining point A to point B and point C to point D to form the cone. Imagine making a party hat, except in graphene only discrete wedge angles are allowed. The *dashed path dakua* circumnavigates the apex (Masako Yudasaka, 2008).

cent, 84 per cent, 60 per cent, 39 per cent, and 19 *per cent* for one, two, three, four and five pentagonal disclinations within an ideal graphene lattice. Zero degrees correspond to six pentagons, *i.e.*, a nanotube. Strong deviations from these discrete angles suggest the formation of wrapped helices rather than continuous cones. In contrast, polygonal rings with more than six carbons induce negative curvature in the graphene sheet. In this context, graphene can be considered as a zero-pentagon carbon nanocone (Charlier, 2004). At a finer level, one may also consider the relative positions of the pentagons within the tip, classifiable by group theory. These pentagonal rings affect both the mechanical and electronic response of the sp2 which demonstrates the orbital configuration of carbon structure, both locally (through an enhanced local density of electronic states near the pentagonal rings) and globally (through new modes of mechanical response and geometrical phases associated with electron paths that circle the apex). (Masako Yudasaka, 2008).

Single wall nanocones (SWNCs) are produced by CO_2 laser ablation of graphite at room temperature in Argon (Ar) gas at about 760Torr with a laser intensity of about 2 kW/cm^2. The production system has been steadily improved, recently reaching a production rate of 1 kg/day with purity higher than 90 per cent. The main impurities are graphitic particles with sizes on the order of micrometers, plus trace amounts of C60. The growth mechanism of SWNCs has been studied with *in situ* optical measurements, and the mechanism is considered as follows. The carbon clusters in the laser plume collide during propagation and form carbon droplets, which might have sizes of about 70nm for the 100-nm dahlia aggregate, if we assume a carbon-droplet density of about 2 g/cm^3. The carbon in the droplets transforms into graphene sheets as the temperature decreases, and each graphene sheet spontaneously rolls up to form a tubule structure, as also seen in a theoretical calculation. The structures of individual SWNCs and the aggregate sizes are influenced by the type and pressure of the ambient gas used in the laser ablation. When SWNCs are produced at a low Argon pressure, the conical portion is shortened and the tubule shapes become thin and bumpy as is illustrated in Figure 3.3. The same change is apparent when the gas

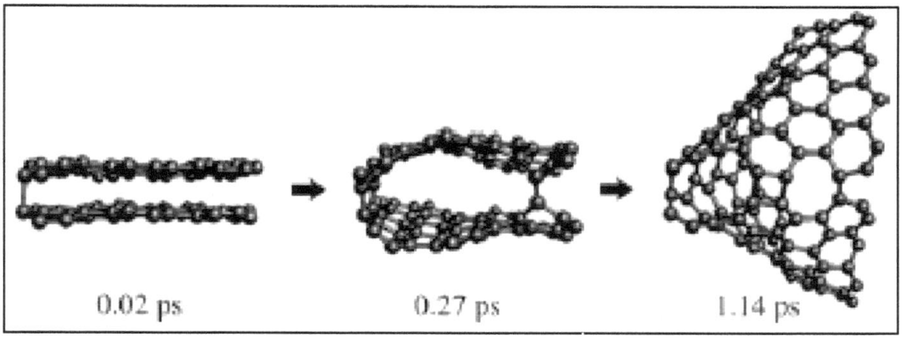

Figure 3.3: Tight-Binding Molecular Dynamics Simulation of the Reactive Collision of Two Graphene Sheets, Suggesting that they can Spontaneously Roll Up to Form a SWNC Tip Structure. The initial distance between graphene layers is small, 0.23 nm, and the temperature in the simulation is 2000 K. The simulation periods are shown below each structure (Masako Yudasaka, 2008).

is changed from Argon to Helium (He) or Nitrogen (N_2). The graphitization of the CO_2 laser ablation products is enhanced by an increase in the ambient pressure, and graphite balls are produced when the Argon gas pressure reaches about 4 atmosphere (Masako Yudasaka, 2008).

When the graphite target includes metal catalysts such as Ni– Fe (Nickel-Iron), the CO_2 laser ablation produces single wall nanotubes (SWNTs) together with SWNCs without heating the system. Production methods not using the CO_2 laser ablation have also been reported, such as an arc-discharge method with carbon electrodes under various conditions (Masako Yudasaka, 2008).

A SWNC is a closed tube in the as-grown state. The holes are easily opened by heat treatment in oxygen gas or oxidative acids. The holes are visible with transmission electron microscopy (TEM) and the dominant hole diameters are 1–1.5 nm when the holes are opened by treatment in oxygen gas at 570 per cent C for 10 min. The size and number of holes are changed by the oxidation conditions.

The functional groups located at the holes edges are also controllable. An interesting property of these holes is that those at the tips of the SWNC can be closed by heat treatment at 1200 per cent C in an Argon atmosphere (Masako Yudasaka, 2008).

This paper is a survey into thermal conductivity properties of carbon nanocone materials to be used in re-entry vehicle by using molecular dynamics simulation. Today in aerospace fields, thermal conductivity of structures for re-entry vehicle is an important subject and in this review paper carbon nanocone materials are introduced and also this paper is a survey into thermal conductivity properties of carbon nanocone materials to be used in re-entry vehicle by using molecular dynamics simulation. Today, carbon nanocone due to its unique properties has many applications in various industries including aerospace industries. One of the interesting and unique properties of carbon nanocone is lower conductivity properties of these materials than carbon nanotubes in higher temperatures. In this paper by molecular dynamics simulations, conductivity properties of carbon nanocone are obtained (Masako Yudasaka, 2008).

Review to Molecular Dynamics Simulations

$$E_{steric\ energy} = E_{str} + E_{bend} + E_{improper} + E_{tor} + E_{vdW} + E_{qq} \quad (1)$$

Molecular dynamics simulation consists of the numerical, step-by-step, solution of the classical equations of motion. At first, we must know that why the molecular simulation is important. MD is one of the methods but there are other methods like Monte Carlo (MC) that can be used. MD gives you dynamics. Other methods can give you the ensemble, but MD gives you a movie. Dynamics are important because biological systems are compartmentalized and are far from equilibrium. In this method at first we must calculate all of potential energies of these atomic structures that is given by (Michael P. Allen, 2004).

E_{str} represents the energy required to stretch or compress a covalent bond that is illustrated in Figure 3.4.

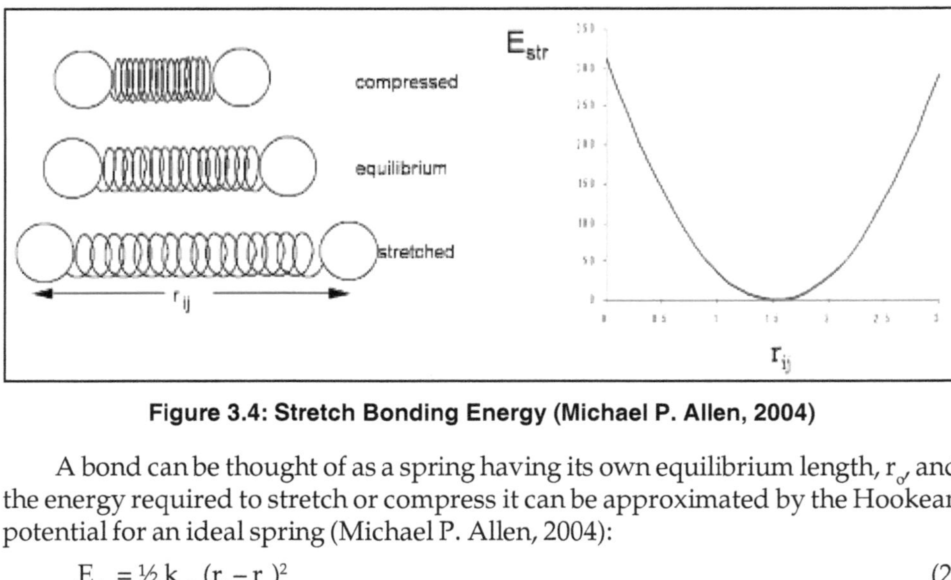

Figure 3.4: Stretch Bonding Energy (Michael P. Allen, 2004)

A bond can be thought of as a spring having its own equilibrium length, r_o, and the energy required to stretch or compress it can be approximated by the Hookean potential for an ideal spring (Michael P. Allen, 2004):

$$E_{str} = \tfrac{1}{2} k_{s,ij} (r_{ij} - r_o)^2 \quad (2)$$

E_{bend} is the energy required to bend a bond from its equilibrium angle, θ_o as is illustrated in Figure 3.5.

Again this system can be modeled by a spring, and the energy is given by the Hookean potential with respect to angle (Michael P. Allen, 2004):

$$E_{bend} = \tfrac{1}{2} k_{b,ijk} (\theta_{ijk} - \theta_o)^2 \quad (3)$$

$E_{improper}$ is the energy required to deform a planar group of atoms from its equilibrium angle, w_o, usually equal to zero that is shown in Figure 3.6.

Again this system can be modeled by a spring, and the energy is given by the Hookean potential with respect to planar angle (Michael P. Allen, 2004):

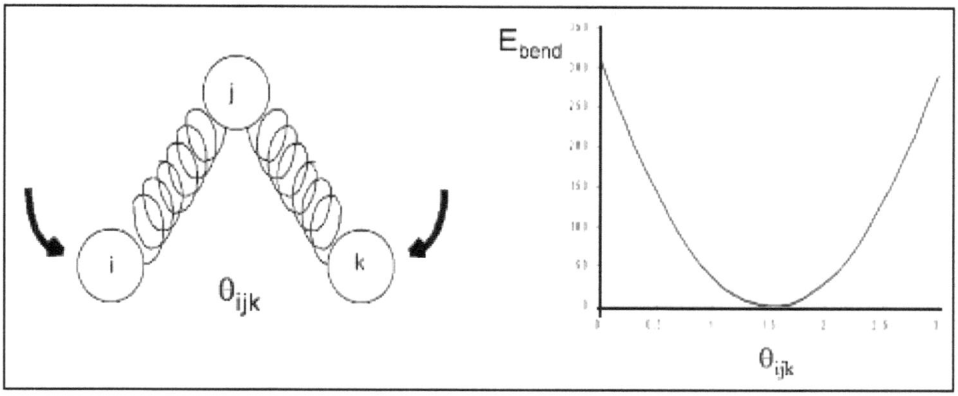

Figure 3.5: Bend Bonding Energy (Michael P. Allen, 2004)

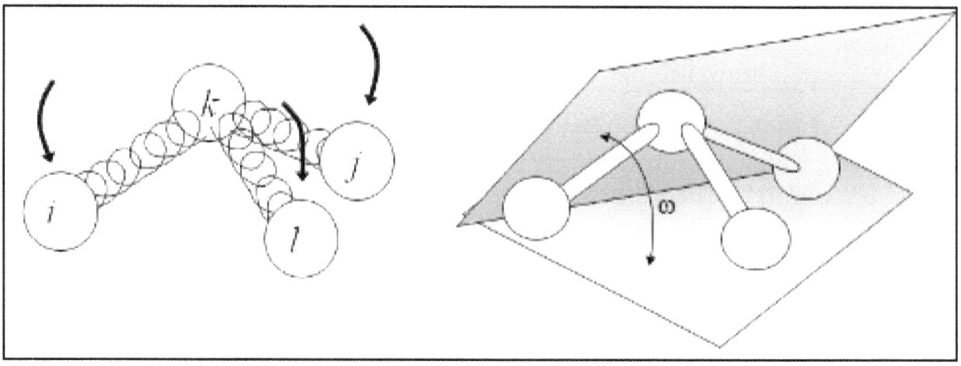

Figure 3.6: Improper Bonding Energy (Michael P. Allen, 2004)

$$E_{improper} = \tfrac{1}{2}\, k_{o,ijkl}\, (\omega_{ijkl} - \omega_o)^2 \tag{4}$$

E_{tor} is the energy of torsion needed to rotate about bonds. Torsion interactions are modeled by the potential equation and are given by (Michael P. Allen, 2004):

$$E_{tor} = \tfrac{1}{2}\, k_{tor,1}\, (1 - \cos\varphi) + \tfrac{1}{2}\, k_{tor,2}\, (1 - \cos 2\varphi) + \tfrac{1}{2}\, k_{tor,3}\, (1 - \cos 3\varphi) \tag{5}$$

In this equation, parameter of can be found from Figure 3.7.

All of these potentials were bonded interaction and at this time we introduce non-bonded interactions (Michael P. Allen, 2004).

E_{VDW} is the steric exclusion and long-range attraction energy (Quantum Mechanics) that it's variation is shown in Figure 8.

Lenard-Jones equation for van der Waals interaction is given by (Michael P. Allen, 2004):

$$E_{VDW}(R) = 4\varepsilon_{ij}\left[\left(\frac{\sigma_{ij}}{r_{ij}}\right)^{12} - \left(\frac{\sigma_{ij}}{r_{ij}}\right)^{6}\right] \tag{6}$$

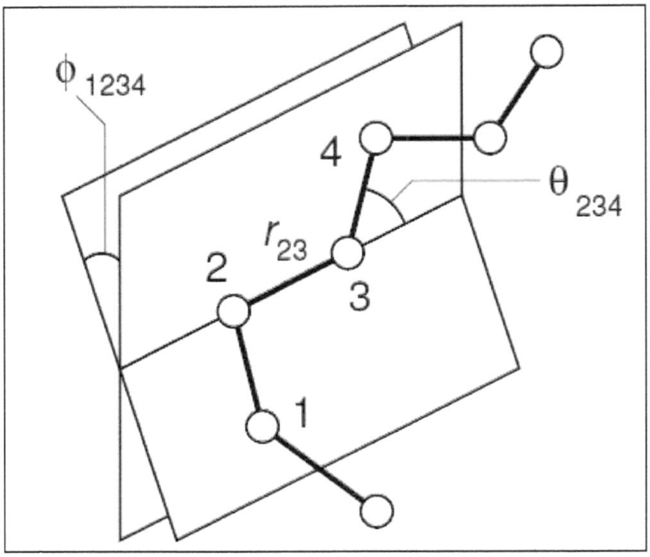

Figure 3.7: Torsion Bonding Energy (Michael P. Allen, 2004)

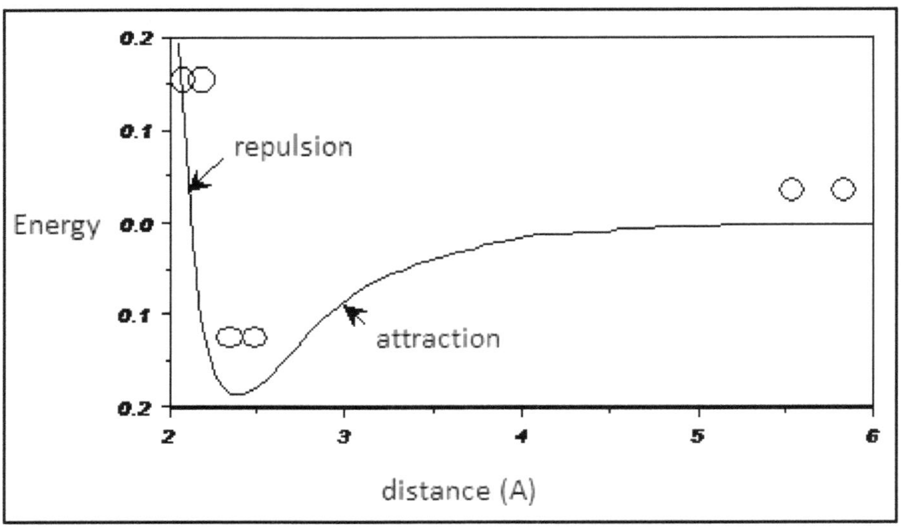

Figure 3.8: van der Waals Interaction (Michael P. Allen, 2004)

With two parameters: σ and ∊ are functions of material properties and they will be calculated from Schrodinger equations.

E_{qq} is the Coulomb potential function for electrostatic interactions of charges and is given by (Michael P. Allen, 2004):

$$E_{qq} = \left(\frac{k q_i q_j}{4 \pi \varepsilon_0 r_{ij}} \right) \tag{7}$$

The q_i and q_j are the partial atomic charges for atoms i and j separated by a distance r_{ij}., ε_0 is the relative dielectric constant. For gas phase calculations ε_0 is normally set to 1. Larger values of ε_0 are used to approximate the dielectric effect of intervening solute (ε_0 ~ 60-80) or solvent atoms in solution. k is a units conversion constant; for kcal/mol, k=2086.4.

So according to these potentials we have:

$$E=\sum_{bonds}k_b\left(r-r_0\right)^2+\sum_{angles}k_\theta\left(\theta-\theta_0\right)^2+\sum_{dihedrals}\sum_{n=1}^{N}K_\varphi^{(n)}\left[1+\cos\left(n\varphi-\gamma\right)\right]+ \\ \sum_{impropers}K_\omega\left(\omega-\omega_0\right)^2+\sum_{i,j}4\varepsilon_{ij}\left[\left(\frac{\sigma_{ij}}{r_{ij}}\right)^{12}-\left(\frac{\sigma_{ij}}{r_{ij}}\right)^{6}\right]+\sum_{i,j}\left(\frac{kq_iq_j}{4\pi\varepsilon_0r_{ij}}\right) \tag{8}$$

Now according to these potentials we can calculate force field (Michael P. Allen, 2004):

$$F_i=-\frac{\partial E_i}{\partial r_i} \tag{9}$$

And according to Newton's law we have:

$$F_i=m_ia \tag{10}$$

By using Verlet's numeric integration method (Michael P. Allen, 2004) and according to the Taylor expansion we have:

$$r(t+\delta t)=r(t)+v(t)\delta t+\frac{1}{2}a(t)\delta t^2$$

$$r(t-\delta t)=r(t)-v(t)\delta t+\frac{1}{2}a(t)\delta t^2 \tag{11}$$

Also we have:

$$r(t+\delta t)=2r(t)-r(t-\delta t)+a(t)\delta t^2 \tag{12}$$

By substituting these equations into equation 10 and solving ordinary differential equation by finite difference method, parameters r(t) and velocity can be calculated.

Thermal Conductivity Properties of Carbon Nanocone Materials

In review, we focus on the cone with the cone angle of 113 degrees, which is the largest angle observed experimentally and theoretically. In our simulations, classical non-equilibrium molecular dynamics method is adopted. The potential energy for carbon nanocone is a Morse bond and a harmonic cosine angle for bonding interaction, which include both two-body and three-body potential terms and is given by (Nuo Yang, 2008):

$$E(r_{ij},\theta_{ijk}) = K_{cr}(e^{-\gamma(r_{\theta}-r_0)} - 1)^2 + K_{c\theta}\frac{(\cos\theta_{ijk} - \cos\theta_0)^2}{2} \tag{13}$$

Where θ_{ijk} represents bending angles and r_{ij} represents distances between bonded atoms. K_{cr} = 478.9, $K_{c\theta}$ = 562.2 are the force constants of the stretch and bend potentials, respectively. The equilibrium carbon-carbon bond length, r_0, is 1.418 Å and the bond angle, θ_0 =120, which are the corresponding reference geometry parameters for graphene. The force field potential used in the current manuscript was developed by Y. Quo, N. Karasawa and W. A. Goddard III by fitting experimental lattice parameters, elastic constants and phonon frequencies for graphite to interpret various properties for fullerene (including vibrational spectroscopy, crystal structure analysis). The accuracy of Morse potential is demonstrated as the calculated thermal conductivity of (10, 10) single wall carbon nanotube with 40 nm in length is 315 W/mK with Morse potential, which is very close to values ~350 W/mK, ~310 W/mK, and ~215 W/mK reported by others (Nuo Yang, 2008).

First, the heat flux in nanocone with forty layers (N=40) is explored, whose length along the side face is around 5.6 nm. Here, fixed boundary conditions are used, that the atoms in the two end layers are frozen. In order to establish a temperature gradient along the longitudinal direction, the nanocone is coupled with Nosé-Hoover heat bathes at the *second and N-1*'th layers, with temperatures are T_{top} and T_{bottom} respectively. The Nosé-Hoover is a deterministic method used in molecular dynamics to keep the temperature around an average. It was originally introduced by Nosé and developed further by Hoover. The heat bath is made into an integral part of the system by adding an artificial variable associated with an artificial mass. Besides Nosé-Hoover heat bath, other methods can also be used to produce a temperature gradient and calculate the heat flux, such as the method developed by Müller-Plathe in which a cold and a hot region of the system are created by switching the velocity of the hottest atom in the cold region with the velocity of the coldest atom in the hot region. Nevertheless, it has been demonstrated that the thermal conductivity and heat flux calculated by MD do not depend on the details of the heat bath used. For instance, the thermal conductivities of silicon nano wires calculated with Nosé-Hoover heat bath agree very well with those calculated with Langevin heat bath. In this paper, all the calculations are done with the same Nosé-Hoover heat bath parameter (thermostat response time) on the two ends (R de Paiva, 2007).

Then the velocity Verlet algorithm is used to integrate the differential equations of motions. In general, the temperature, T_{MD}, is calculated from the kinetic energy of atoms according to the Boltzmann distribution that is given by (Nuo Yang, 2008):

$$Kinetic\ Energy = \sum_{1}^{N}\frac{1}{2}mv_i^2 = \frac{3}{2}NK_BT_{MD} \tag{14}$$

Where E is the mean kinetic energy, v_i is the velocity of atom, m is the atomic mass, N is the number of particles in the system, and K_B is the Boltzmann constant. It is worth pointing out that this equation is valid only at very high temperature ($T>>TD$, TD is the Debye temperature). Debye model gives a good approximation for the low temperature heat capacity of insulating, crystalline solids where other contributions (such as highly mobile conduction electrons) are negligible. In this case, the Debye model can only be said to approximate the lattice *contribution* to the specific heat. When the system average temperature is lower than the Debye temperature, it is necessary to apply a quantum correction to the MD calculated temperature. The difference between the MD calculated temperature and quantum corrected temperature depends on the Debye temperature. However, the Debye temperature of carbon nanocone is not known. Although, many theoretical studies have been conducted on the Debye temperature of carbon nanotube, the results have been controversial. For instance, 473K, 475K, 580K, and 1000K are reported in different literatures. The concerns of the current paper are the heat flux and rectification effects (a related change of heat flux) which do not depend on the accurate value of temperature-but the temperature difference. So, we don't do the quantum correction to T_{MD} in this work (Nuo Yang, 2008).

The heat flux J_l along the nanocone is defined as the energy transported along the surface in unit time. The expression of flux is given by:

$$J_l(t)=\sum_i v_{i,l}\,\varepsilon_i+\frac{1}{2}\sum_{ij} r_{ij,l}(\vec{F}_{ij}.\vec{v}_i)+\sum_{ijk} r_{ij,l}(\vec{F}_j(ijk).\vec{v}_j) \tag{15}$$

Where ε_i is local site energy, F_{ij} is two-body force, and $F_j(ijk)$ is three-body force and force field is given by:

$$F_i=-\frac{\partial E_i}{\partial r_i}=-\frac{\partial\left(K_{cr}(e^{-\gamma(r_\theta-r_0)}-1)^2+K_{c\theta}\dfrac{(\cos\theta_{ijk}-\cos\theta_0)^2}{2}\right)}{\partial r_i} \tag{16}$$

Also we have:

$$\cos\theta_{ijk}=(r_{ij}.r_{ij})^{-\frac{1}{2}}(r_{jk}.r_{jk})^{-\frac{1}{2}}(r_{ij}.r_{jk}) \tag{17}$$

After substituting equation 17 into 16, we can calculate force field and then we can substitute it in equation 15.

Simulations are performed long enough such that the system reaches a stationary state when the local heat flux reaches a constant along the nanocone. All results given in this work are obtained by averaging about 4×10^6 time steps. The time step is set as 0.4 fs. The total thermal conductivity contains contributions from electrons and phonons. However, phonons dominate the heat transport in carbon nanocone because it is a semiconductor (Nuo Yang, 2008).

In physics, a phonon is a collective excitation in a periodic, elastic arrangement of atoms or molecules in condensed matter, such as solids and some liquids. It represents an excited state in the quantum mechanical quantization of the modes of vibrations of elastic structures of interacting particles. Phonons play a major role in many of the physical properties of solids, including a material's thermal and electrical conductivities. Hence, the study of phonons is an important part of solid state physics. A phonon is a quantum mechanical description of a special type of vibration motion, in which a lattice uniformly oscillates at the same frequency. In classical mechanics, this is known as the normal mode. The normal mode is important because any arbitrary lattice vibration can be considered as a superposition of these *elementary* vibrations (Fourier analysis). While normal modes are wave-like phenomena in classical mechanics, they have particle like properties in the wave particle duality of quantum mechanics (Nuo Yang, 2008).

In this paper, we set the temperature of top as $T_{top} = T_0(1-\Delta)$ and that of bottom as $T_{bottom} = T_0(1+\Delta)$. Where T_0 is the average temperature, and Δ is the normalized temperature difference between the two ends. Therefore, the bottom of nanocone is at a higher temperature when $0<\Delta$, and the top has a higher temperature when $0>\Delta$. Also the nanocone behaves as a good thermal conductor under positive thermal bias and as a poor thermal conductor under negative thermal bias so in this lecture we will consider J_- and J_+ that are the heat current from top to bottom corresponds to $\Delta<0$ and current from bottom to top corresponds to $\Delta>0$ respectively. Also, we can introduce heat flux percent:

$$heat\ flux\ ratio = \frac{J_+ - J_-}{J_-} \qquad (18)$$

Discussion

The heat flux per cent from top to bottom and opposite versus T_0 when T_0 differs from 200K to 1000K is shown in Figure 3.9, here our nanocon's length is 9nm. As the

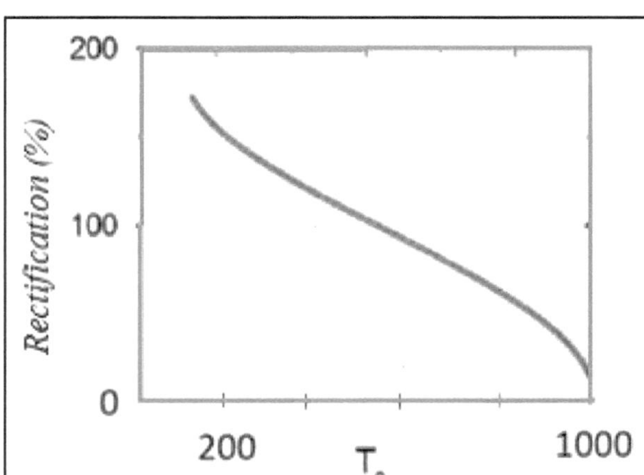

Figure 3.9: Parameter of Heat Flux Per cent from Top to Bottom and Opposite Versus T_0 when T_0 differs from 200K to 1000K

temperature increases, the heat fluxes increase, and the increase of " J is higher than the increase of + J, which leads to the reduction of the rectification ratio. According to molecular dynamics simulation, we studied carbon nanocone near the temperature of 1000 degrees Kelvin and we couldn't use these results for re-entry phase that re-entry vehicle will face a temperature near 1500 degrees Kelvin but carbon nanocone can be used as an insulator to protect our subsystems in the phase that our vehicle is orbiting in a specific orbit and the vehicle thermal environment can vary considerably, depending upon a variety of naturally occurring effects since in this stage maximum temperature will be about 300 degrees Kelvin and our results are reliable for this stage.

So for carbon nanocone materials in the temperature of 1000K we have:

$$\mathrm{Re}\,ctification = \frac{J_+ - J_-}{J_-} \times 100\% \approx 0 \tag{19}$$

Finally by molecular dynamics simulations, it can be found that heat flux percent of these advanced materials have nothing to do with the geometric parameter of length. This result can be seen in Figure 3.10 that heat flux percent from top to bottom and opposite versus length of nanocone is drawn.

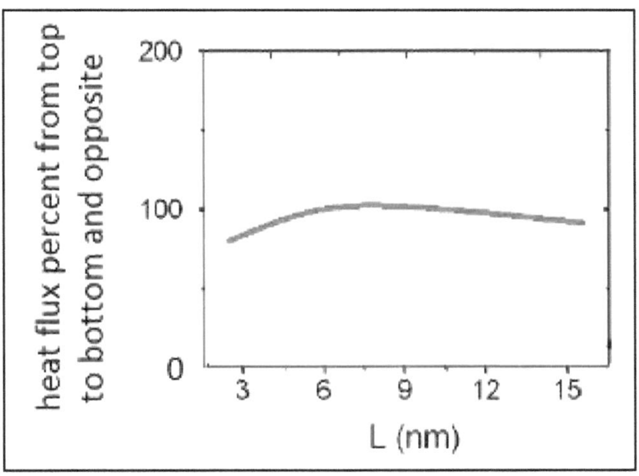

Figure 3.10: Parameter of Heat Flux Per cent from Top to Bottom and Opposite versus Length of Nanocone with the Temperature of 1000 K

Conclusions

This paper was a survey into thermal conductivity properties of carbon nanocone materials to be used in re-entry vehicle by using molecular dynamics simulation. Today in aerospace fields, thermal conductivity of structures for re-entry vehicle is an important subject and in this paper carbon nanocone materials are introduced to be used in these structures and also this paper was a survey into thermal conductivity properties of carbon nanocone materials to be used in re-entry vehicle by using molecular dynamics simulation. Today, carbon nanocone due to its unique properties has many applications in various industries including aerospace industries. One of the interesting and unique properties of carbon nanocone is lower conductivity properties of these materials than carbon nanotubes in higher temperatures. In this paper by molecular dynamics simulations, conductivity properties of carbon nanocone were obtained. By molecular dynamics simulations, it can be found that conductivity thermal properties of these advanced materials have nothing to do with the geometric parameter of length. Also if initial temperature in re-entry conditions increases, thermal conductivity will be reduced for these advanced materials. So for the structures of re-entry vehicles that must tolerate high initial temperatures, application of these advanced materials is very vital because in such temperatures, thermal conductivity property of these materials is very low and as insulation materials, they can be used in the structures of re-entry vehicles. In this paper we studied carbon nanocone materials with temperature of 1000K for using in re-entry vehicles but for higher temperatures like 1500 degrees Kelvin in re-entry phase, molecular dynamics doesn't have any response for this simulation, so we must use some ceramics like Aluminum Nitride nanocone instead of carbon nanocone in such temperatures.

Aluminum nitride (AlN) nanocone is a ceramic material with good dielectric properties and high thermal conductivity, when compared to other ceramics. It has a thermal expansion coefficient comparable to that of Si and is non-reactive with the usual semiconductor process chemicals and gases (R de Paiva, 2007).

Hence, in order to fully exploit these materials for innovative nano devices, the structural and electronic properties of these nanostructures need to be investigated in more detail. In this paper, we investigated thermal conductivity properties of carbon nanocone by molecular dynamics simulation and in future works we can investigate thermal conductivity properties of Aluminum Nitride Nanocone by other methods.

Acknowledgements

We would like to acknowledge and extend our heartfelt gratitude to the following persons who have made the presentation of this Lecture paper possible:

Members and Staff of Iranian Research Organization for Science and Technology and NAM S&T Centre for their vital support.

Members of National Elite Foundation faculty for their support.

Most especially to our family.

And to God, who made all things possible.

References

Allen, Michael P., 2004. Introduction to molecular dynamics simulation.

Charlier, 2004. Graphene as a zero–Pentagon carbon nanocone.

Paiva, R de, Azevedo, S. and Kaschny, J.R., 2007. Structural and electronic properties of aluminium nitride nanocones.

Xi, Wei Jing, 2007. Simulation of carbon nanocones using molecular dynamics.

Yang, Nuo, Zhang, Gang and Baowen, L., 2008. Carbon nanocone: A promising thermal rectifier, Singapore.

Yudasaka, Masako, Iijima, Sumio and Crespi, Vincent H., 2008. Single wall carbon nanohorns and nanocones.

Chapter 4

Production of Nano Silver Water Suspension Using Colloidal Silver Generator

S.J. Beden, H.I. Dambus**, R.I. Atiyah, O.N. Abbod, M.R. Mohameed, J.F. Hamoodi, H.A. Ahmad, W.H. Mahdi and H.A Hasan*

Ministry of Science and Technology
Directorate of Advanced Material Researches
Iraq

ABSTRACT

Silver is a powerful natural antibiotic noble metal which has been used for thousands of years with no harmful side effect since we have observed. Today colloidal silver is rapidly gaining favor in the medical community and water treatment reflecting a kind of revival in the public health community sector. In this research, we designed colloidal silver (CS) generator using two rods of silver connected to pulse DC high power supply to get arc discharge caused to evaporate silver rods which are immersed in water. Later, we made improvement on this design by using Direct Current power supply with high purity silver. Rods are arranged in certain arrangement then we get many silver water samples according to different parameters. In order to determine physical properties of (CS) solutions, Atomic force microscopy (AFM), UV- Visible spectrometry, Atomic absorbtion spectrometry (AAS), and electrical conductivity were applied to each samples. The results show that (CS) generator is effective tool to get narrow range nano- colloidal silver water suspension with size

E-mail: *sab_salim2001@yahoo.com; **h-dombus@yahoo.com

distribution (40–560) nm, color of each solution is characterized to particle size silver particle/ ionic solution. The electrical conductivity of sample depended on the ionic/particle silver concentrations which were about (4.3-34.6) ppm with pH values (5.81-10.05)

Keywords: *Colloidal silver, Water suspension, Environment.*

Introduction

Colloidal silver (CS) is a natural antibiotic which has been used through out the world for centuries as a mean to destroy microbes of all kinds and to correct many health problems [1,2]. The use of colloidal silver is spreading rapidly throughout the medical community in the healing of burned victims. Silver acts as a second immune system for all humans by destroying bacteria and viruses of all kinds. It is toxic to *bacteria, viruses, yeasts, fungi* (*molds*), *protozoa* and parasites in the egg stage[3,4]. (CS) will destroy *Staph* and *Strep* bacteria which are so common in health problems. Silver does not attack bacteria directly, but rather decomposes certain enzymes than aerobic bacteria viruses, *yeast,* and *molds* require the silver to act as a catalyst and is not consumed in the process. It is probable that this indirect action is also the reason for bacteria can not develop a resistance to silver, as they do to is antibiotic[5,6].(CS) are extremely minute silver particles which suspend with particle size at rang (4-560nm) in water, with positive electric charge. The smaller the particles, the more effective (CS) it has proven to be.The best colloid silver is produced at a molecular level. A small DC current is passed thought an electrolytic medium (distilled water) with silver electrodes. Minute, molecular size particles are drawn of positive electrode, having positive electrical charge. This electrical charge is of primary importance to healing and antibacterial qualities [7–9]. The charge slowly dissipates, especially when exposed to light. Since our blood is also a colloid, the harmonious way the colloids enter the body may well make colloidal silver, the safest medicine one earth [10–12].

In principle, current - techniques for nanometer sized particles preparation can be divided into two categories: physical and chemical. [13].This paper describes three physical methodology of silver water suspension; first method (A) by electro polishing method, (B) arc-discharge method [14], and (C) electrolysis method[15–19].

Experimental

Materials

Silver wires (99.99) purity, 1-1.5 mm in diameter, distilled water (PH 5.1), DC pulsar power supply, magnetic stirrer.

Methods of Prepairing Colloidal Silver (CS)

Electro Polishing Technique

First method, we start with the idea of electro polishing process using Ag wire immersed in electro solution as recorded in handbook of materials [20]. Ag wire used as a node surrounded by Al foil acts a cathode immersed in polishing solution (200ml H_2O, 300ml C_2H_2OH, 400ml H_3PO_4) as shown in Figure 4.1. DC power supply applied

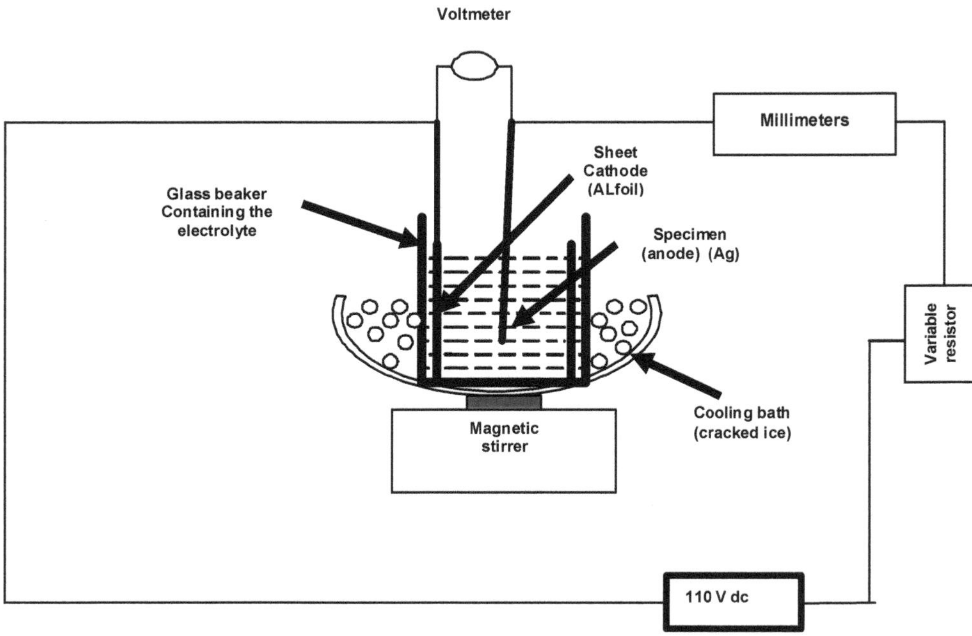

Figure 4.1: Ag Electropolishing Circuit where Ag Rods as Cathode and Anode Immersed in Polishing Solution

a potential difference in the range (0.2–10) V for 1 hr. Five samples prepared were characterized white to grey color as shown in Figure 4.2. The (CS) produced in this method show high adhesion degree related to phosphorus acid, so we made chemical - physical separation process to separate (CS) from phosphoric compounds as shown in Figure 4.3.

Arc-Discharge Technique

The DC arc-discharge system as in Figure 4.4 consist of DC pulser power supply provides a pulse voltage of (70–100) V for 2 ms and (20–40) V around 10 µs. The electrical current could reach during that moment (4–10 A) where the etching current caused arc-discharge pulse etched Ag wires (anode and cathode). Ag wire evaporated and condensed in water. Servo control system keep the gap between the Ag electrodes in few microns. Other parameters should be controlled during this technique such as the speed of stirring, time of process, water volume, distance of the gap, arc-discharge current and other parameters. The transparent solution converts to characteristic pale yellow color as in Figure 4.6.

Electrolysis Technique

By using DC power supply (30 V) and two Ag high purity rods immersed in distilled water as shown in Figure 4.7, we get (CS) simply, cheaply faster than other methods. We prepared (CS) samples in different applied voltage (10–30) V for 1 hr. The transparent solution characterized with different degree of white- yellow colour as shown in Figure 4.8.

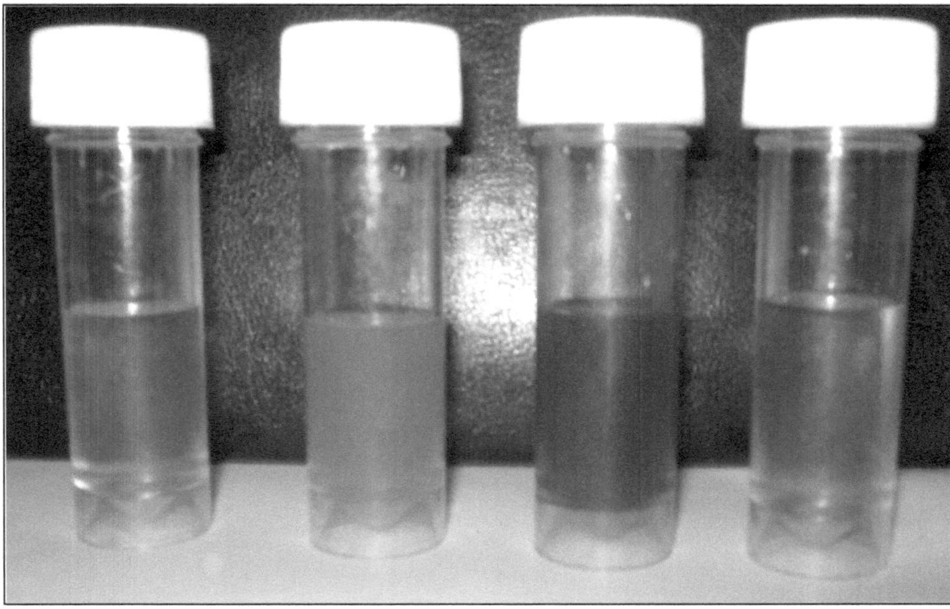

Figure 4.2: Nano-Colloidal Ag (Nag) Samples (White, Yellow, Grey Colour)

Figure 4.3: Separation Process to Separate NAg from Phosphoric Compounds

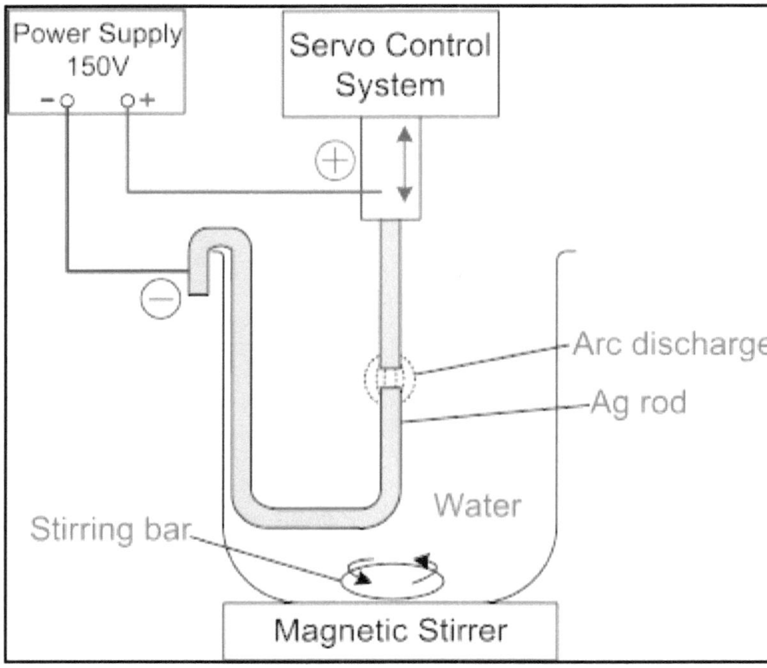

Figure 4.4: The Arc-Discharge System for Nano-Colloidal Ag Preparation Process

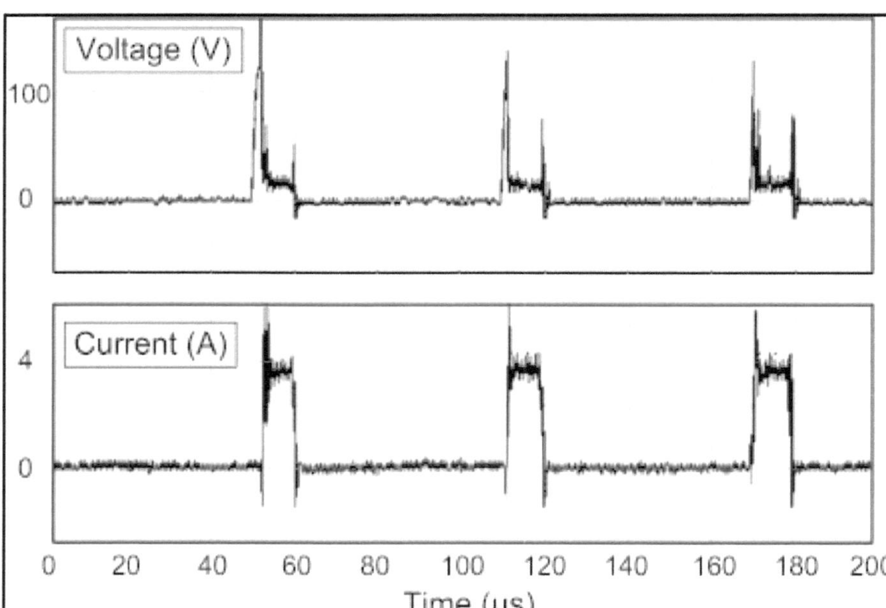

Figure 4.5: Current-Voltage Pulse during Arc-Discharge Periods

Figure 4.6: Samples of Colloidal Ag Water Suspensions (Pale Yellow Colour) Related to different Particle Size

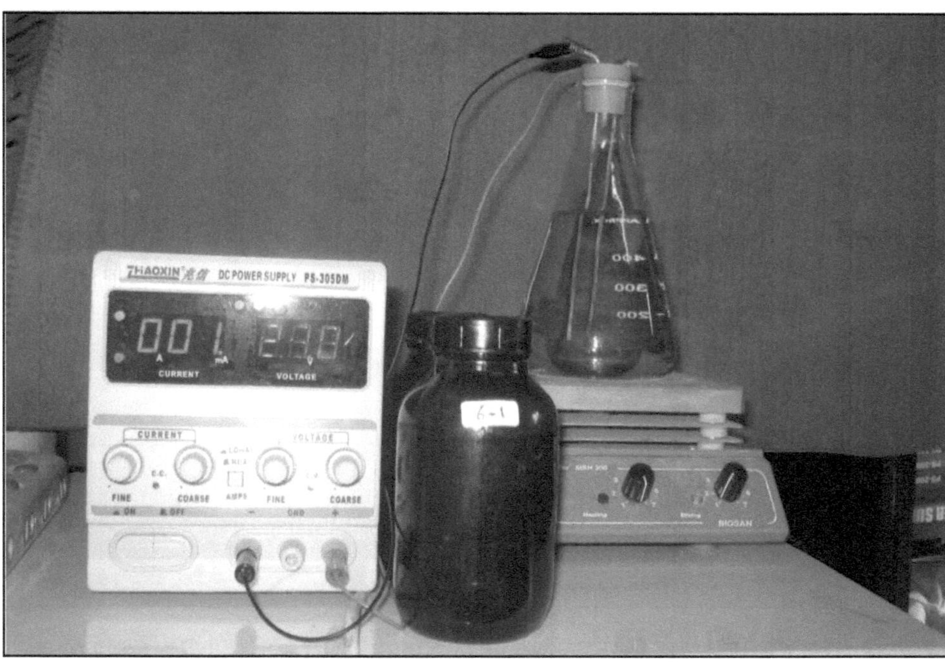

Figure 4.7: The–Electrolysis Arrangement Technique for Preparation Colloidal Silver Suspension

Figure 4.8: Samples of CS Water Suspension by Electrolysis Technique Different Voltage for 1hr (Pale-yellow colour)

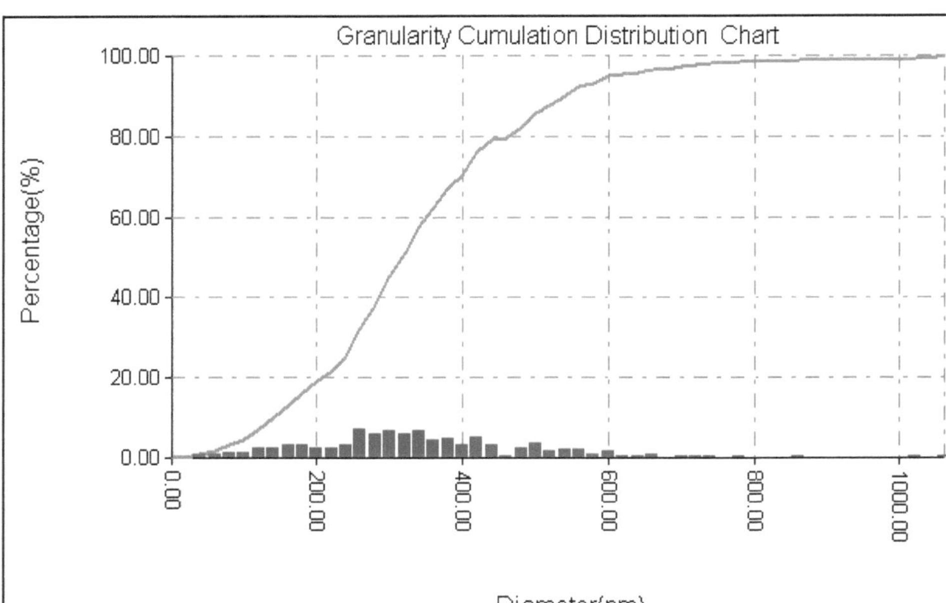

Figure 4.9a: Particle Size Distribution of CS Prepared by Electrolysis Technique

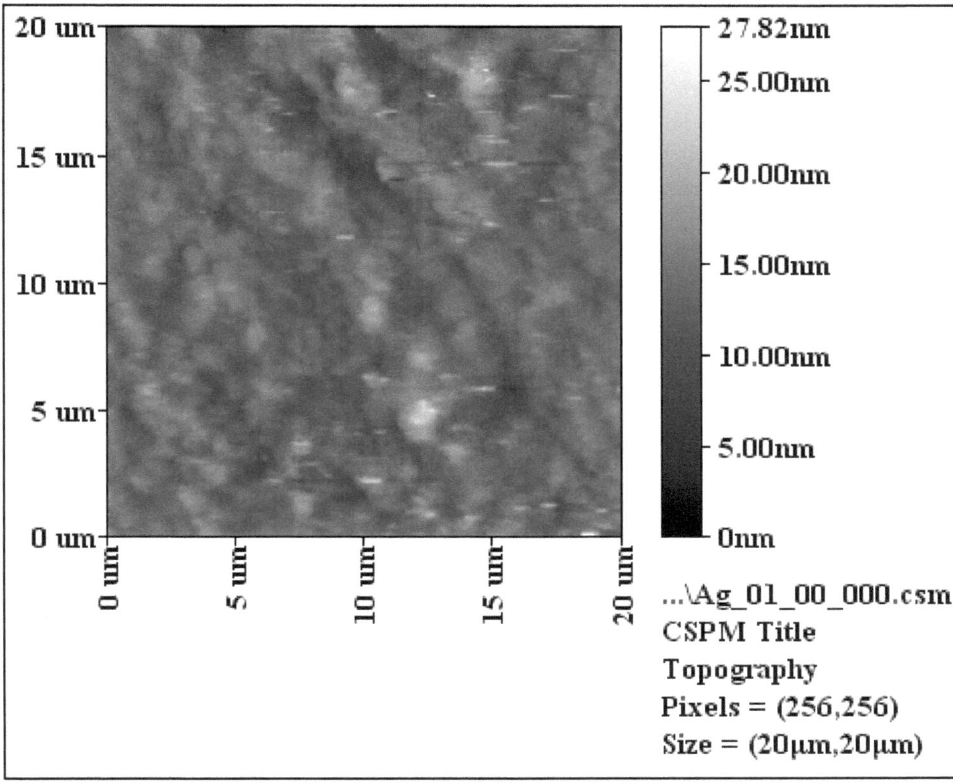

Figure 4.9b: Surface Morphology of CS Films by Electro Polishing Technique

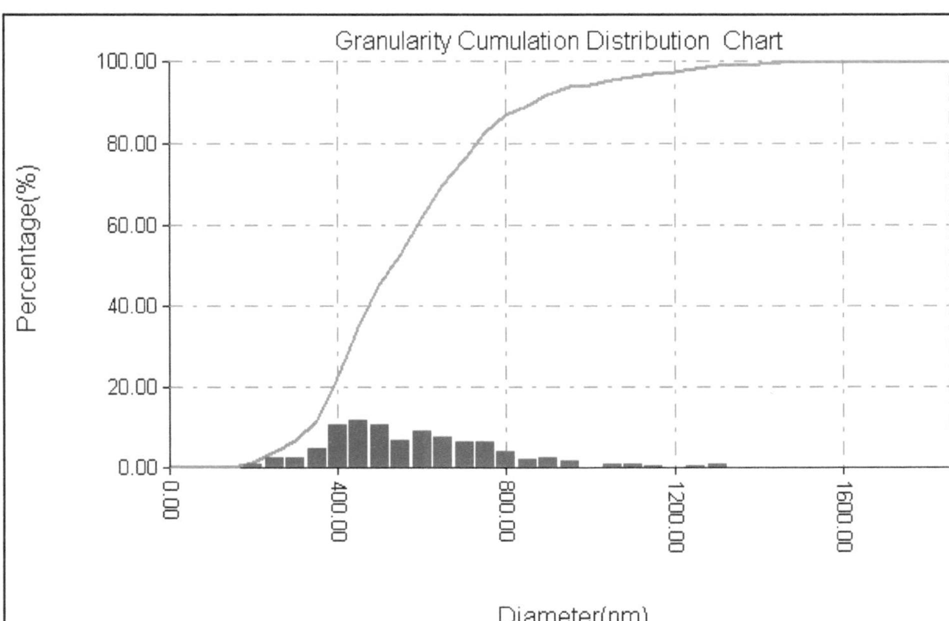

Figure 4.10a: Size Distribution of CS by Arc-Discharge Technique

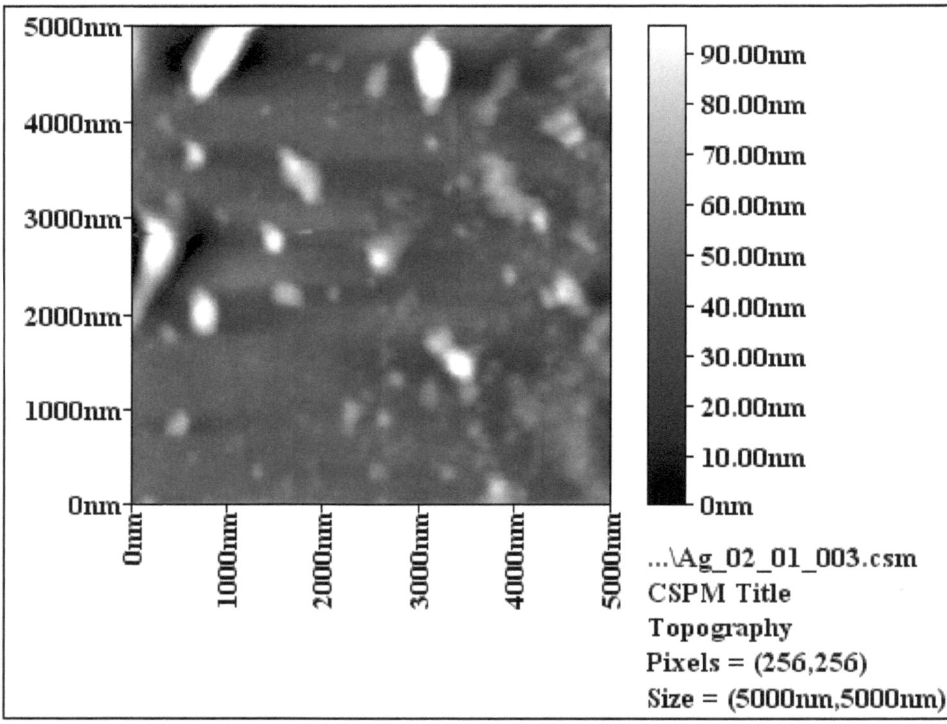

Figure 4.10b: AFM Image of CS Film by Arc-Discharge Technique

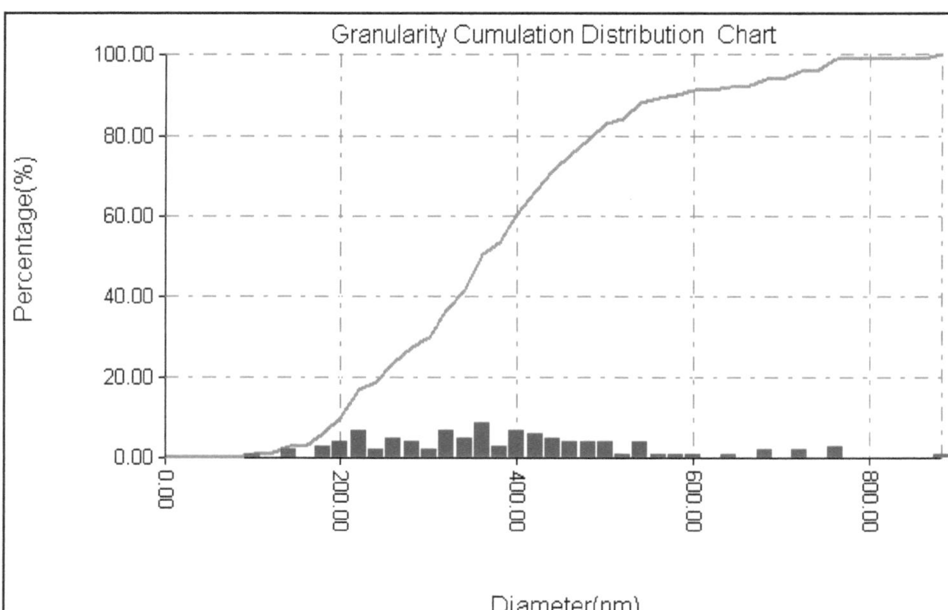

Figure 4.11a: Size Distribution CS by Electrolysis Technique

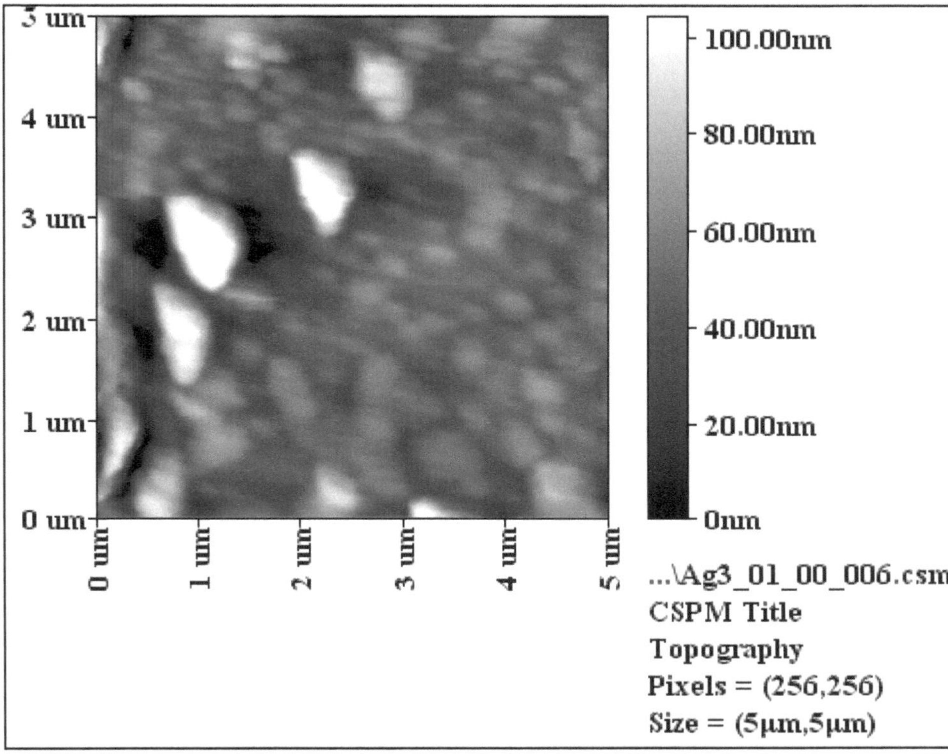

Figure 4.11b: AFM Image of CS by Electrolysis Technique

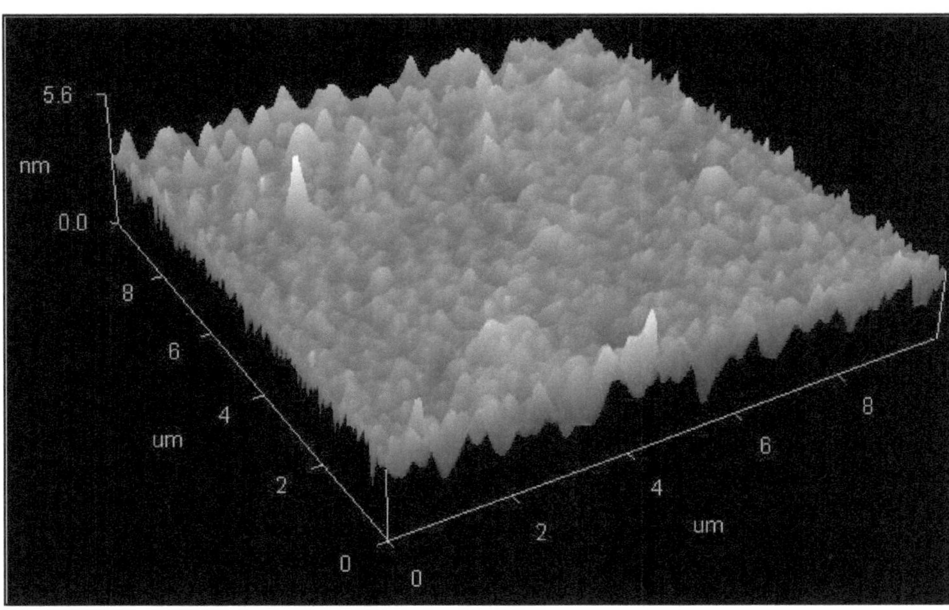

Figure 4.12: The 3D Image of Ag Nano-Colloidal Suspension Spread on Glass Substrate

Characterization of Ag Nano Colloidal (CS)

The Ag colloid solutions were examined optically using UV-Visible spectrophotometer (type shamedzo 20). Silver solutions were sprayed on glass substrates for size distribution determination by AFM (Angstrom dvanced-typeAA3000)), Electrical conductivity (COM 83–Radiometer Copenhagen), PH, Ag concentration by AAS (AA Spectrophotometer phoenix -986) were also examined for each sample.

Results and Discussion

Figures 4.9a,b shows the AFM image of spherical Ag particles (method A) with size distribution about (40-500) nm and Figures 4.10a,b shows AFM image of spherical Ag particles (method B) with size distribution about (100 - 360) nm and Figures 4.11a,b AFM image of Ag particles (method C) prism with size distribution about (100-560) nm, Figure 4.12 show the 3D image of Ag nano-colloidal suspension spread on glass substrate.

Table 4.1 show the chemical value of CS suspension water prepared by Electrolysis method.The table show the effect of applied voltage during preparation stages on the concentration and electrical conductivity increased proportionally and the variation of Ag concentration with ph.[16].

Table 4.1: Characterization of CS Samples Prepared by Electrolysis Technique

Sample Code	*Concentration ppm*	*pH*	*Electrical Conductivity μs/cm*
1	4.3	5.81	2.99
2	10	6.64	3.07
3	22.1	6.73	3.35
4	22.8	6.82	4.85
5	32.2	7.18	9.97
6	34.6	10.05	28.10

The effects of light and temperature on colloidal silver were observed since the agglomeration process is detected by UV-vis spectrophotometer and color appearance; so we concentrate on right preparation parameters and storage conditions of silver samples. During silver nano particles production by (B,C) method, water decomposition (*e.g.* electrolysis) was observed where gaseous hydrogen and oxygen appear in the water medium. Hydrogen and oxygen start to interact with newly prepared silver nano particles. Since Hydrogen (molecular or atomic forms) does not adsorb on silver particle surface (method B, C) at room temperature[21], and also is not dissolved in water. It is ultimately removed from water suspension to gas phase while oxygen (especially atomic) could adsorb and react with silver surface at room temperature, Ag cluster could easily create bonds with water particles in water environment (method B, C). Finally, negatively or positively charged Ag narrow rang nano particles can be created in water as shown in the model Figure 4.13.

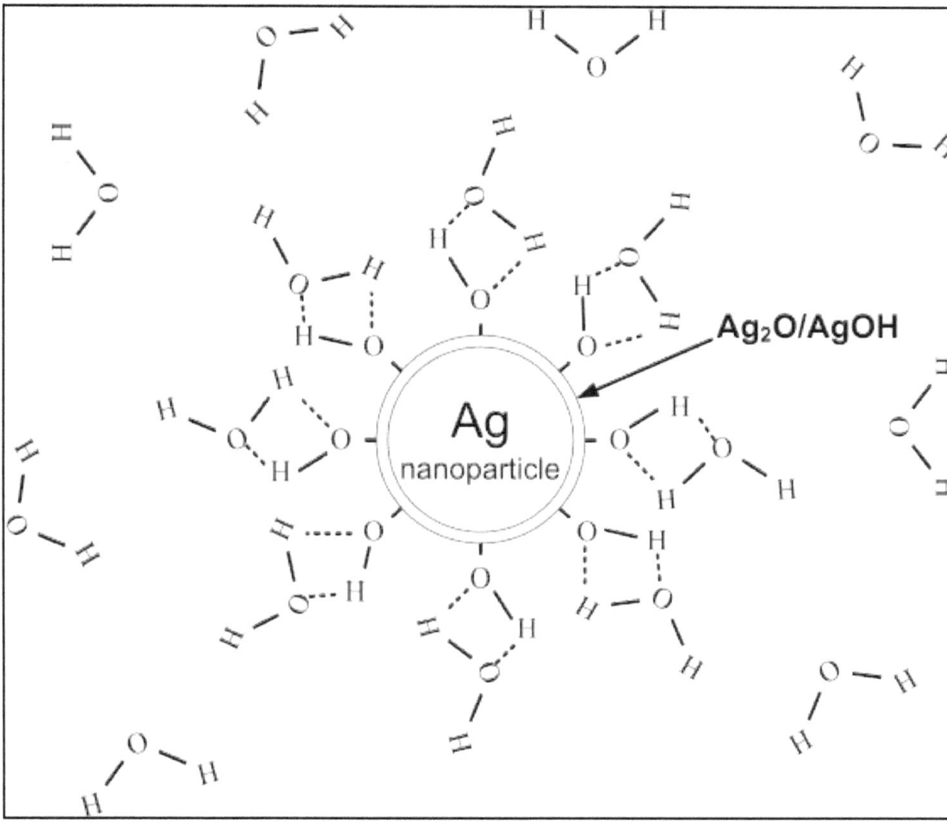

Figure 4.13: Model Scheme of Charged Ag Particles Suspended in Liquid Medium (Distilled water)

Figure 4.14 shows a typical UV-visible absorption spectra for the silver colloid suspension and this results show maximum absorbance in some band of spectra at range about 420nm according to three techniques [18,19].

Figure 4.15 shows the x-ray diffraction diagram of the silver samples in order to study their structure and it shows the crystalline pattern with characterized peak position 2θ (33.0337) degree with high relative intensity (100 per cent) to plane (111) and second peak position 2θ (43.9817) degree with relative intensity (52 per cent) to plane (200) and these results were in agreement with other studies [14].

The Ag nano narrow range particles (CS) due to hydrogen bonding were suspended in water without any stabilizers or surfactants. Positively charged particles (method C) Ag nano particles surrounded by water molecules having (4-34.6) ppm and electrical conductivity about (2.99-28.10) µs/cm with pH (5.81-10.05).Ag (CS) (method B) could be heated to 100 °C without affecting its structures[20,21]. The DC method (B,C) based on pure components, e,g metal rods and deionizer water, seems to be a promising alternative for metal noble metal such as nano gold and nano titanium since the high surface - area to volume ratio of nano particles can create their

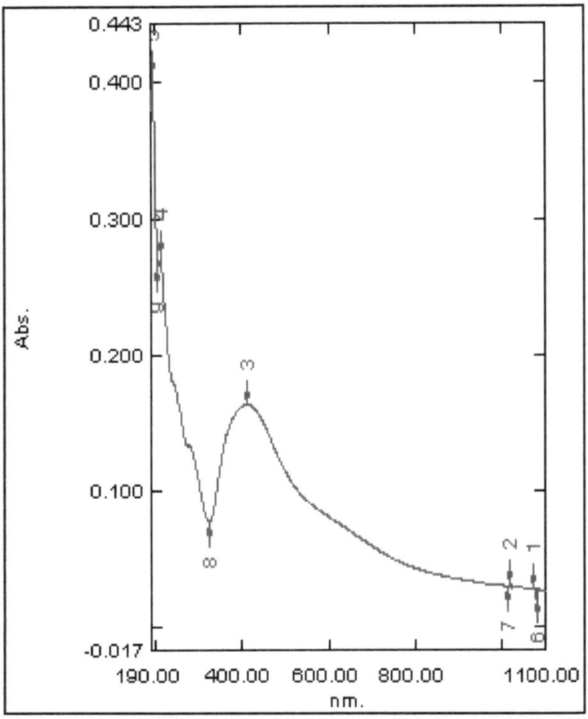

Figure 4.14: UV-vis Spectra of Colloidal Silver Suspension

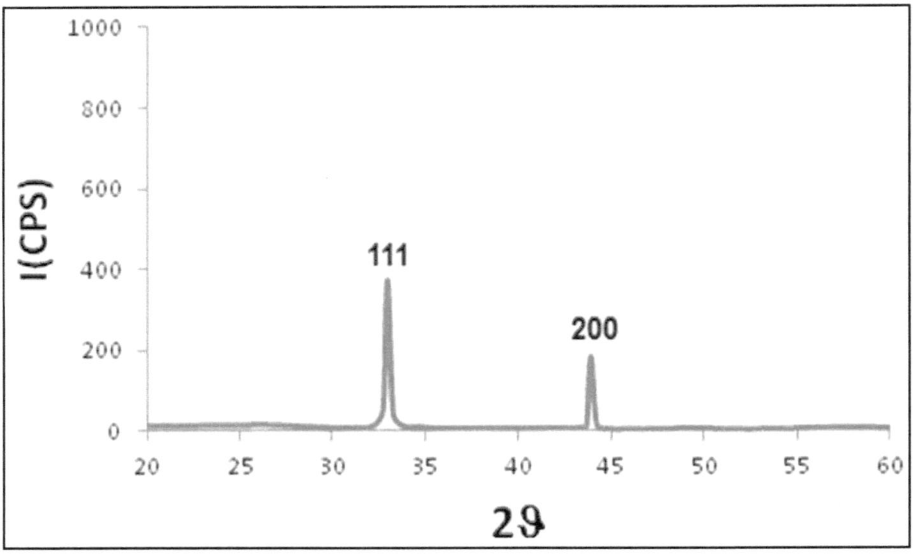

Figure 4.15: XRD Diagram of Nano Silver Thin Film

unique physical chemical, mechanical, and quantum size effected proprieties[22] that lead to their potential application such as catalysis for plants, animals and human [23,24].

The (CS) could be followed by the filtration process and centrifugation technique many times in different ways to get dominate nano size particles with low cost clean environmental process [25].

Conclusions

1. The current DC method in pure water to prepare (CS) has been successfully developed for narrow range nano particles Ag production.
2. The present work prepares low cost effective scientific route to form nano silver powders by controlling (I,V,T, and Ph,)parameters.
3. Colloidal silver solution has a characteristic UV-vis spectrometry with UV band at resonance peak of about 420nm.
5. According to AFM and OM ; silver particle solution has semi spherical shape with size distribution 40-560 nm.
6. The silver nano particles less than 50 nm could be obtained using arc discharge technique followed by filtration and centrifugation process.
7. Silver nano particles are formed in water has characteristic X-ray diffraction peak at plane (111 and 200).

References

1. Zhao, G. and Sterens, Jr. S.E., 1998. Multiple parameters for the comprehensive evalution of succeptibility *E. coli* to the silver ion. *Biometarls*, 11: 26–33.
2. Tisun and Seff, K., 1994. *Chem. Rev.*, 4: 857.
3. Hung, H.H., Xp, Ni and Glilloy C.H. *et al.*, 1996. *Langmuir* 12: 909.
4. Wisnhoven, S.W.P., Pejburg, W.G.M. and Herberis, C.A.A. *et al.*, 2009. *Nanotechnology*, 3(2): 109–138.
5. Fumo, F., Morly, K.S., Wong, B., Sharp, B.L. and Amold, P.L. *et al.*, 2004. Silver nanoparticles and polymeric medical devices: A new approach to prevention of infection. *J. Antimicrob. Chemother.*, 54: 19–24.
6. Hanonda, T., Mye, A., Donovan, B., Shih, A., Reater, J.D. *et al.*, 2000. A novel surfactant nanoemulsion with auniqe non-irritant topical antimicrobial activity agansit bacteria enveloped viruses and fungi. *Microbiol. Res.*, 156: 1–7.
7. Sondi, I. and Salopek-Sondi, B., 2004. Silver nanoparticles as antimicrobial agent: A case study on *E. coli* as amodel for gram-negative bacteria. *J. Colloid Interface Sci.*, 275: 177–182.
8. Sterns, Z.G., Jr. S.E., 1998. *Biometa.*, 511: 27–32.
9. Aymonier, A.U., Schloher, Beck, L. and Antonieo, *et al.*, 2002. *Chem. Commun.*, 24: 309, 18–9.
10. Mallikariuva, K., Narastmha, G. and Pillip, G. *et al.*
11. Hong, K., Lpark, L.,Sal, I.H. and Kang, T., 2006. *Polymer Physics*, 44: 2468–2474.
12. Kmt L. Xiat, Lenaghan S.C., Zhang.

13. Jones, S.A., Bowler, P.G., Walker, H. and Parons, D., 2004. Controlling wound bioburned with a novel silver-containing Hydrofiber dressing. *Wourd Repair Regen.*, 12(3): 288–294.

14. Tien, D.-C., Liao, C.-Y., Hang, J.-C. and Lung, J.-K. *et al.*, 2008. Novel techniqes for preparing nano-silver water suspension by the arc-discharge methods. *Rev. Adv. Mater. Sci.*, 18: 750–756.

15. Khanna, P.K., Singh, N., Kulkarni, P., Deshmukh, S., Charan, S. and Adhyapk, P.V., 2007. *Material Lett.*, B: 336–337.

16. Tsung, T.T., Chang, H., Chen, L.C., Han, L.L., Lo, H. and Lin, M.K., 2003. *Mater. Tran.*, The Japan Institute of Metals, 44, 1138.

17. Gaidan, L., Pziiea, A., Crobann, C. and Marti, T., 2009. *Biotechnological Lett.*, 14(5): 4465–4472.

18. Luoma, S.N., 2008. Silver nanotechnologies and the environment project on emerging nano technologies . The Raw Charitable Trust, Renis.

19. Aslan, K., Leonenko, Z., Lakowicz, J.R. and Geddes C.O., 2005. *T. Phys. Chem. B.*, 109: 3157–3162.

20. *Metal Handbook* Vol. 9: *Metallography and Microstructure*, 1985, Table 1, Fig.6, p. 50.

21. Dus, R. and Nowicka, E., 2003. *Progress in Science*, 74: 39.

22. Lok, C.N., HO, C.N., Chen, R. and He, Q.Y. *et.al.*, 2007. Silver nanoparticles: Partial oxidation and antibacterial activities. *J. Biol. Inorg. Chem.*, 12: 527–534.

23. Pal, S.,Tak, Y.K. and Song, J.M., 2007. Does the antibacterial activity of silver nanoparticles depend on the shape of the nanoparticles? A study of the Gram-negative bacteria *E. coli. Appl. Environ. Microbiol.*, 73: 1712–1720.

24. Murata, T., Kanao, K., Oshikawa, M. and Takamtsu, T., 2005. Effect of Pb, Sb, In and Ag contamination on the proliferation of soil. *Water, Air, Soil Pollut.*, 164: 103–118.

25. World Health Organization (WHO), 2008. Silver and silver compounds: Environmental aspects. From the website: http://www.who.int/ipcs/publications/cicad/en/cicad44.pdf.

Chapter 5

Kenya Institutional Policy Framework for the Development and Utilization of New and Advanced Raw Materials and a Report on the Preliminary Year 2008 Country Survey on Utilization of Raw Materials

E.K. Mwangi[1], Moses Rugut[1] and Crispus Kiamba[2]*

[1]Department of Research Management and Development, P.O. Box 30568–00100, Nairobi, Kenya

[2]Permanent Secretary, Ministry of Higher Education Science and Technology, Kenya

ABSTRACT

Kenya is rich with a diverse of natural resources and fauna which are major sources of Raw Materials.

There are five key areas within the industrial sector where raw materials are heavily utilized, namely; agro-industrial, chemicals and pharmaceuticals, energy, ceramic and building materials, engineering and metal works. Other major strategic industries are in food, textile and leather.

* *Corresponding Author:* E-mail: emwangi23@yahoo.com

In that respect, there are a number of Government Sectoral Ministry Institutions handling aspects of utilization of raw materials and development of new ones as part of their mandate. To support these MOHEST has in the STI policy proposed a framework to for development and utilization of new and advanced raw materials.

On the other hand, as an initial step to implement the initiatives on raw material utilization proposed in the STI policy, in 2008, the Kenya Ministry of Higher Education, Science and Technology (MOHEST) got engaged with Raw Materials Development Centre (RMRDC)–Nigeria in a Bilateral Cooperation.

Kenya has previously participated in the Action Committee on Raw Materials ARCM initiatives previously but there had not been a serious engagement as the current one established through MOHEST –RMRDC cooperation. Since then the Ministry was exposed to ACRM initiative in more details.

As a follow up, a rapid survey of the utilization of Raw materials in Kenya was done in (2008) using questionnaires and information data collection sheets sent by the RMRDC–Nigeria.

This paper is presented in two parts; a) Kenya Policy strategies on utilisation Raw Materials and; and b) Results of a rapid raw material utilisation survey carried out in 2008 under the MOHEST-RMRDC partnership.

Keywords: *Kenya Vision 2030, Innovation, Raw materials utilization, Industrialization.*

Introduction

In Part I, this paper presents the key highlights on the various sectors utilizing Raw Materials in Kenya and justifies the need for the development of a defined policy framework on the utilization of raw Materials issues in Kenya for socio-economic development.In Part II - it reports on a of a rapid raw material survey carried out in Kenya between February and June in 2008.

Part I: Policy Framework for the Development and Utilization of New and Advanced Raw Materials in Kenya

The Kenya Government has been implementing the Economic Recovery Strategy for Wealth and Employment Creation (ERSWEC) since 2003, as its medium term plan to address development challenges of economic growth and poverty alleviation. In its long term development strategy, outlined in Vision 2030, the Government of Kenya envisages a nation that is globally competitive and prosperous with a high quality of life by the year 2030 (Kenya Vision 2030, 2007). Science, Technology and Innovation (ST&I) has been identified as a foundation for Vision 2030. In pursuit of the vision, Science, Technology and Innovation will be harnessed to stimulate technological and industrial transformation that will lead to sustained economic growth of 10 per cent per annum, and social well-being in the next 25 years.

The Government therefore, formulated the National Science, Technology and Innovation (ST&I) Policy and Strategy (Kenya STI-Policy 2007-2012; Kenya Ministry of Higher Education Science and Technology Medium Term Plan for 2008–2012) to guide and promote placed on identified National Priority Growth Sectors that have high potential to harness ST&I in attaining the targeted 10 per cent annual economic growth. Utilisation of raw materials for industrialization is highlighted in some of these major National Priority Growth Sectors.

In that respect, there are a number of Government Sectoral Ministry Institutions handling aspects of utilization of raw materials, while others are incorporating this activity as part of their mandate (*e.g.*, the Kenya Industrial Research and Development Institute (KIRDI) Strategic Plan 2008–2012).

Policy Strategies

In line with the above, the government has formulated the following S&T strategies (Republic of Kenya Ministry of Higher Education Science and Technology Medium Term Plan for 2008–201200 with respect to raw materials;

Indigenous Resources and Traditional Knowledge (IRTK)

The objective of this policy strategy is to exploit the full potential of science, technology and innovation through generation, protection, preservation, evaluation, updating, adding value to, and utilizing the extensive indigenous resources and traditional knowledge, as applicable to national development. This will be accomplished through evaluating and updating of the extensive indigenous resources and traditional knowledge and promoting partnership between the Government, Private sector, NGO's-and other Stakeholders. This will enhance utilization of IRTK results for national development,that are Kenyan in origin and character.

Initiatives and Flagship Projects Focusing on Raw Material Utilization to Implement the Kenya Vision 2030

In the Kenya Vision 2030, several initiatives and flagship projects that address the sectors utilizing raw material have been identified as follows, under the "New and Emerging Technologies" programme among others;

Under the Vision 2030, the following Innovation Sectors are closely linked to or dependent and are meant to promote raw material utilization as one of the enabling sectors for social-economic development;

1. Industrialization
2. Manufacturing
3. Value addition/natural resources
4. Infrastructure
5. Indigenous knowledge–raw materials–herbal medicines, etc, water purification, food preservation
6. Knowledge based economy (science, mathematics, engineering, ICT)

The above innovation sectors are addressed through the following"New and Emerging Technologies" initiatives and flagship projects;

New and Emerging Technologies Programme

The programme will focus on developing new scientific understanding and capabilities in emerging fields of ST&I. New scientific and technological opportunities in areas identified by researchers will be developed. Assessment of new discoveries or newly-observed phenomena which could indicate risks or problems to society will be undertaken. Finally, initiatives focusing on specific, highly challenging objectives in emerging scientific and technological fields, and involving groups of complementary projects will be undertaken. The programme will aim at supporting visionary research with the potential to open new fields for ST&I.

Building and Construction Materials Project

The aim of this project is the provision of cost effective high quality materials for high-tech infrastructure and transport designs.

Nanotechnology Project

The aim of this project is to develop appropriate high strength materials which are low cost, effective and efficient. Nanotechnology allows the manipulation of properties of materials at the molecular or atomic level, giving rise to products with enhanced properties, faster production processes, lower production costs, much smaller manufacturing equipment, cleaner environment or new manufacturing systems. Nanoelectronics are expected to create an impact on the local Electrical and Electronics industry. An example is the production of nanoscale devices for semiconductors. Other novel techniques include production of nanoparticles and carbon nanotubes. When these new methodologies are adopted in production, it will result in devices and systems with new levels of performance. The advantage of nanotechnology is its ability to be applied in diverse industrial use like automobile industry, textile industry, health services among others.

Laser Technology Project

The project's objective is to develop and promote laser technology for efficient and rapid industrial transformation. Laser technology has a wide application in areas of manufacturing, consumer electronics, telecommunications, data communication, surveying and construction, medicine and the military among others.

Metal Processing Technologies Project

The histories of the developed and the newly developed economies of the world show that the establishment of material processing industries has, nearly always, had the greatest contribution to the growth of their national wealth. Industrialization has been the direct result of the need for readily available, affordable, adequate and high quality goods and services. Electronic consumer goods, textiles, silverware, industrial machinery, defense machinery, automobiles etc, are examples of end products of certain manufacturing processes. The great demand for these products all over the world is a statement of a large amount of the physical substances used as inputs in their production or manufacturing. In other words, a great demand for 'semi-processed materials'.

The aim of this project is to improve technologies for processing materials for the improvement of the appearance properties of the semi-processed material or to produce new materials and to provide the technologies required for expansion and diversification of the metal industry. The project will also expand capacity for training in metal casting including the development of training programmes at the Craft and Artisan levels. This should enable more utilization of metals available in the country including scrap metal of which much is being exported.

Biotechnology Initiative

The objective of this project is to develop and promote biotechnology for efficient and rapid industrial, agricultural and health transformation. The biotechnology project will ensure the development of biomaterials, including polymers, metals, ceramics and composites, which are produced synthetically or biologically, for use in the efficient treatment and management of diseases with growth potential in bio-generics and molecular diagnostics for diseases, as well as vaccines for tropical diseases. The project will also ensure the development of biosensors, which represent a powerful new evolutionary field in analytical measurement technology. The commercial application of biosensors is in the areas of healthcare, industrial process control, agriculture, veterinary medicine, defense and environmental pollution monitoring.

Biopolymers, which are biologically produced polymers with unique functional uses, such as stabilizers, thickeners, gallants, binders, dispersants, lubricants, adhesives and drug delivery agents. The main beneficiary of biopolymers is the food processing industry. Other industries which will benefit include cosmetics, pharmaceuticals, packaging, paper and textiles. Advances in plant genetics and fermentation and purification technologies can eventually influence a shift towards the application of more biopolymers than synthetic petroleum-based polymers. The rise in crude oil prices will provide opportunities for the country to develop alternative sources of fuel, for example, biodiesel. The project offers opportunities to partner with industry and establish Public Private Partnerships as well as promote essential foreign investment to add value to Kenyan biotechnology and in developing key regional and international markets.

Value Addition Initiative

The aim of the project is to develop and promote value addition and agro-processing technologies to ensure high value exports through efficient and effective production. The project will;

1. Ensure low cost production and processing, high productivity and quality of agricultural produce and products as well as encourage growth of rural non-farm enterprises.
2. Ensure the exploitation of indigenous foods for consumption and commercialization and will attract public-private partnership arrangements.
3. Ensure value addition to farm waste as well as assess and advice on the capabilities and capacities of meteorological services vis a vis the demands in the agricultural sector.
4. Reduce supply side inefficiencies and improve market access
5. Develop skills in agro-processing.

Health Initiative

One of the activities under this initiative will be focus on ensure commercial production of traditional plants for medicinal use and streamlining of the intersectoral and public-private sector partnerships in hi-tech medical technologies and improvement of cost effectiveness and efficiency of the health care delivery system.

Development of a National Mineral Industry Project: - The aim of this project is to build Capacity and capabilities for mineral resource exploration and exploitation, establish and implement Resource based human conflict mechanisms in the mining zones and determine the national mineral resource base. In addition, it will ensure efficiency in mining and value addition in mineral processing.

In view of the above described policy strategies, one of the objectives of Ministry of Higher Education, Science and Technology (MOHEST) - Kenya participation in this "*International Conference on Harnessing the Potentials of New and Advanced materials for Developing Economies*" which was held at Abuja, Nigeria–was to share experiences with the partners from the various NAM S&T Centre countries with respect to policy matters.

Part II: Results of a Rapid Preliminary Raw Material Utilization Survey Carried Out in 2008 in Kenya

Kenya MOHEST– Raw Materials Research and Development Council (RMRDC),– Abuja collaboration started in February 2008 through the Action Committee on Raw Material initiative.

The Action Committee on Raw Materials (ACRM) of Group of Seventy-Seven and China (G-77) was conceived as a platform to strengthen co-operation among developing countries, especially members of the G-77, for the purpose of developing raw materials for industries and for other value added uses. ACRM is therefore, one of the major initiatives of the G-77 and China under its Economic Co-operation among Developing Countries (ECDC) programmes. The ACRM was established at the sixth meeting of the Inter-Governmental Follow-up and Coordination Committee (IFCC) of the G-77, held in Havana, Cuba in September, 1987.

The overall objective of the ACRM, when it was conceived is to strengthen cooperation among developing countries especially in the G-77, for the purpose of developing raw materials in their primary and processed forms.

The Raw Materials Research and Development Council (RMRDC), a Parastatal under the Federal Ministry of Science and Technology of Nigeria is serving as the International Secretariat of the ACRM (RMRDC– Annual Report–2006).

The work reported here was aimed at collecting preliminary data from Kenya on Raw Materials utilization to add to Raw Materials information system at the ACRM.

Material and Methods

As a first step, information gathering forms (data questionnaires) were sent to Kenya from RMRDC, Nigeria. A preliminary survey on utilization of Raw materials done between February and June –2008 in Kenya by MOHEST. The Institutions and industries in major industrial towns in Kenyan visited or from which information was gathered included among others;

- ☆ Pharmacy and Poisons Board
- ☆ Kenya Industrial Research Institute
- ☆ Kenya Agricultural Research Institute
- ☆ Mines and Geology Department
- ☆ Mining companies
- ☆ Agricultural processors
- ☆ Textile industries
- ☆ Food processors
- ☆ Kenya Medical Research Institute
- ☆ Sugar mills
- ☆ Paper mills
- ☆ Private Sector–(manufacturing)
- ☆ SME's and informal sector
- ☆ International investors (*e.g.* in minerals)
- ☆ Research organizations and Academic institutions

In summary, the key details collected during the information collection included;

1. Details on nature and type of organization
2. Key types of Raw Materials required (types, quantities, availability, uses)
3. Sources and suppliers of industrial inputs (local or imported)
4. Degree of processing of primary raw material (local or foreign technology)
5. Uses and types of products
6. Materials exported as raw
7. Problems encountered in processing the raw materials

Results

The results of the analysis of the information and data collected on utilization of Raw materials in the various organizations sampled are analysed and presented in Table 5.1. Broadly, there are five key areas within the Industrial Sector where raw materials are heavily utilized, namely, agro-industrial, chemicals and pharmaceuticals, energy, ceramic and building materials, engineering and metal works. Other major strategic industries are in, food, textile and leather. The data collected corresponded well to the information reported in various Kenya Government - Institutional Annual Reports 2005-2007 (see references).

Discussion

The key outcomes of the survey are discussed and summarized as follows;

1. Most processors were profit driven - they therefore were mainly interested in one line product and occasionally would not invest heavily on other side products.

Table 5.1: A Summary of Currently Commercially Utilized Raw Materials in Kenya, Based on a Preliminary Survey Carried Out from February to June 2008, among the Relevant Organizations and Stakeholders in Various Regions of the Country

Raw Material	*Uses and Utilisation*
Cassava	Starch and its derivatives, modified starches, food
Maize	Animal Feeds, Starch Pectin, Alcohol, Adhesives, Sugars, Oils, Corn Flakes
Tomato/Pawpaw	Purce, Juice, Jam
Banana	Flour, Beverages, Crisps, Jams, Soft Beverages,
Oranges/Pineapple	Fruit Juice Concentrate, Essentials Oils, Flavourings
Mango	Juice, Jam, Jelly, Beverages, Liquor
Coffee	Beverages, coffee husks, charcoal
Sugarcane	Sugar And Derivatives, Molasses
Sorghum	Malt, Animal Feed
Millet	Animal Feeds, Alcohols
Wheat/Barley	Wheat Flour, Bread, Confectionaries,Alcohol, animal feeds
Soyabeans	Soy Oil, Soy Sauce, Soya Bean Meal, Milk
Cashew nut	Food Condiments, Oil Resin, Varnishen,
Cotton	Fibre, Cotton Seed Oil, Cake
Tobacco	Cigars, Cigarettes
Timber	Furniture, Pulp and Paper, E.T.C
Potato (Irish/sweet)	Chips, Starch, Alcohol
Sunflower	Oil, Soaps, Feeds, E.T.C.
Rice	Human food, Bran
Fish	Fish Meat, Animal feeds
Livestock	Meat, Dairy products, hides and skins, bone meal, manure
Cashew	Nuts, Juice, Oil
Coconut	Fibre Mat, Jute Bag, Copra, Oil, Cake, Perfumes, Brooms
Acacia	Tannin
Neem Tree	Insecticides, Local Drums, Gum
Diatomite	Used as absorbent, oil well cementation, pesticide carriers, oil spill absorbent, anticking and acid refining
Fluorspar	Used as a neutral flux by the alcel industry and manufacturer of hydrofluoric acid (HP) which is a basic chemical in the manufacture of refrigerants, aerosols, aluminum fluoride, artificial cryolite, elemental fluorine. Fluorspar is also used by the ceramic industry in glass manufacture, making fiberglass insulation, welding rod coatings, enamels and miscellaneous fluxes
Lime	Used in the steel industry and as a fluxing agent. It is used in the manufacture of chemicals (especially for water treatment) and in buildings and soil stabilization.
Limestone	Used as aggregate in all types of construction and manufacture of cement, lime, flux stone, railroad ballast chemicals, etc

Contd...

Table 5.1–*Contd...*

Raw Material	*Uses and Utilisation*
Phosphate rock	Monthly used for fertilizer production and chemicals
Potash	Fertilizer and chemicals
Salt	Used in the chemical industry especially the chloride and caustic soda sector, agriculture water treatment, metal production.
Soda ash	Used mainly in the glass and chemical industries especially, for sodium-based chemical and detergent in manufacture.
Limestone	Caustic soda manufacture effluent and water treatment

2. Agricultural based manufacturing–Food, fruits, maize meal–are dominant in the market with their by products being utilized into animal feed products and oil products.
3. For mining–there are several multi-international oriented (local branches) companies with parent companies in Europe and Overseas. These mining firms also tend to specialize on one mineral product line which has a defined market.
4. There is still a lot of information on raw material utilization not fully especially in the private sector.
5. There are extensive research and production activities on raw material and over 60 per cent of these are agro-based.
6. The awareness on Raw Material utilization as not as strong as it should be and also; there is fragmented national institutional linkages.
7. While the Government has a variety of institutions utilizing or conduction research on New and advanced Raw Materials, a better or enhanced coordination mechanism is required.

Challenges and Problems of Development and Utilization of Local Raw Materials in Kenya

1. Information release/data exposure was a hindrance. This is because of the norms among manufacturers or utilisers of Raw Materials that their business strategies world leak to competitors.
2. One of the major problems is the lack of data on availability of mineral raw materials. The associated problems are inadequate funds for exploration and exploitation of the minerals and also manpower development.

Conclusions and Recommendations

There is an urgent need to establish a Raw Materials utilization monitoring Unit within the Government (MOHEST) to constantly gather data for comparison with other developing countries. This Unit would act as a Raw Materials information system to link with the Action Committee on Raw Materials (ACRM) for constant updating on the Kenyan Raw Material utilization.

Kenya would also act as a focal point for the Eastern African region to collect information on Raw Material utilization for ACRM.

Acknowledgements

Acknowledgments are made to the following;

1. The RMRDC–Nigeria for encouraging Kenya to actively participate in the ARCM initiative,
2. The Permanent Secretary–MOHEST-Kenya for supporting the rapid raw material utilisation survey carried out in 2008 and;
3. NAM S&T Centre for initiating jointly the workshop with RMRDC and the latter for hosting the event.

References

1. Kenya Industrial Research and Development Institute (KIRDI) Strategic Plan 2008–2012.
2. Ministry of Higher Education, Science and Technology Science, Technology and Innovation Bill (2009).
3. Republic of Kenya Ministry of Higher Education Science and Technology Medium Term Plan for 2008–2012.
4. Republic of Kenya Ministry of Ministry of Higher Education Science and Technology – Strategic Plan 2008–2012.
5. Research, Innovation and Technology Sector (RIT) Medium Term Expenditure Framework 2010/11–2012/13 Technical report (2009).
6. Republic of Kenya Vision 2030 A Globally Competitive and Prosperous Kenya (2007).
7. Republic of Kenya – Institutional Annual Reports–(2007–2009) – from:
   - Kenyan Industrial Research and Development Institute
   - Kenyan Association of Manufacturers Information Centre
   - Ministry of Industry
   - Agricultural Information Center.
   - Kenyan Agricultural Research Institute
   - Mines and Geological Department
   - African Centre for Technology
   - Kenya Bureau of Standardization
   - Kenya National Chamber of Commerce and Industry
8. The Raw Materials Research and Development Council (RMRDC)–Annual Report–2006.

Chapter 6

Synthesis of Zinc Oxide Nanoparticles by Alkaline Precipitation Method

Phyu Phyu Win and Myint Sanda Win*

Nanotechnology Research Department,
Metallurgical Research and Development Centre,
Ela, Leway Township, Naypyitaw
Ministry of Science and Technology, Myanmar

ABSTRACT

The spherical Zinc Oxide nanoparticles were successfully synthesized by alkaline precipitation technique from zinc acetate dihydrate and sodium hydroxide using triethanolamine (TEA) as a surface active agent in order to obtain the particles more spherical. The uniform spherical ZnO nanoparticles with the range from 150 nm to 300 nm were observed from this method. The morphologies of zinc oxide nanoparticles prepared were determined by scanning electron microscope (SEM). The elemental composition of zinc oxide nanoparticles was carried out by SEM attached Energy Dispersive X-ray Diffractometer (EDX) and the crystallization phase structures were studied by X-ray diffraction (XRD).

Keywords: *Zinc oxide nanoparticles, Zinc acetate, Spherical, Precipitation, Triethanolamine.*

Introduction

Recently, the unique properties of nanomaterials have motivated the researchers to develop the simpler and inexpensive techniques to produce nanostructures of

* *Corresponding Author:* E-mail: pphyuwin@gmail.com

technologically important materials (Shah, 2009). Nano- structure zinc oxide, an n-type II-VI compound semiconductor with a wide band gap of 3.3 eV, has attracted considerable more and more attention as a promising candidate for its wide range applications such as varistors, photocatalysis, gas sensors, solar cells, pigment, transducers, etc. High crystallinity of zinc oxide nanoparticles is one of the most important factors in achieving a high UV emission efficiency (Nyffenegger, 1998; Satoshi, 2004; Husnu, 2008; Yan, 2008; Hai, 2004).

Monodispersed ZnO particles with defined morphological characteristics, such as spherical, ellipsoidal, needle, and rod-like shapes have been obtained. ZnO nanoparticles can be prepared on a large scale at low cost by simple solution methods, such as alkali precipitation, thermal decomposition, hydrothermal synthesis, sol-gel methods, microemulsions, spray pyrolysis, template-free method, and other routes. Nanomaterials that are made hydro-thermally are generally high-quality, single crystals with a diversity of shapes and sizes. In this technology, the most significant point to be explored is a rapid and simple method for producing nanometer size particles that emit high luminescence intensity (Hsieh, 2007; Aneesh, 2007).

This study was aimed to develop the solution route synthesis of uniform and spherical ZnO nanoparticles by alkaline precipitation technique. The ZnO nanoparticles were synthesized from zinc acetate dihydrate and sodium hydroxide in the ethanol rich phase using triethanolamine (TEA) as surface acting agent. The crystal structure, morphology and particle size of nanoparticles were investigated.

Experimental Design

Alkaline Precipitation Technique

The chemical reagents used in this work were $Zn(CH_3COO)_2.2H_2O$, NaOH and TEA of analytical grade purity. Zinc acetate solution was prepared by dissolving 0.001 mole of $Zn(CH_3COO)_2.2H_2O$ in 50 ml of water. The alkaline solution was prepared by dissolving NaOH (0.002 mole) and TEA (0.003 mole) in 950 ml of absolute ethanol under stirring until the homogeneous solution. The zinc salt solution was added into the alkaline solution under magnetic stirring at 60-70°C and this solution was continuously heated for different reaction times: 1h and 2h. It was then allowed to cool naturally at room temperature. After the reaction was complete, the resulting white product was washed with ethanol to remove the organic agent TEA and hydroxide ions. The product was filtered and dried in oven at 60°C for three hours.

Characterization of ZnO Nanoparticles

The crystal structure of the synthesized particles was determined by X-ray diffraction (XRD, D8 Advance Bruker Co.,) with Cu Kα radiation λ= 1.5406. The particle size and morphology of the products were investigated by scanning electron microscope (EVO®60 SEM) and the elemental composition was obtained from energy dispersive X-ray diffraction attached to SEM (EDX INCA Penta FETx3 OXFORD Instruments).

Results and Discussion

In the alkaline precipitation of ZnO nanoparticles from zinc acetate and sodium hydroxide using TEA as a surface active agent, the ethanol rich phase was the best dispersing agent and the best solvent. The triethanolamine dissolves in the ethanol. After the ZnO particles have surface active agent TEA to adsorb in the particles surface, a diffuse layer of ethanol- rich liquid around the particles is created. The zinc cation was soluble in water and growth was controlled by the diffusion of Zn^{2+} ions in the ethanol-rich layer. The TEA tended to make the particles more spherical. Without the organic additives or without using alcohol as dispersion medium, rods rather than spheres formed (Hsieh, 2007).

The X-ray diffraction patterns for the ZnO nanoparticles prepared at different reaction times are shown in Figure 6.1. The sharp diffraction peaks of the synthesized products indicated their good crystallinity. Figure 6.1a shows the XRD pattern of the as-prepared ZnO nanoparticles of 1 hour reaction time and Figure 6.1b shows that of 2 hours reaction time. Figure 6.1c shows the XRD pattern of the ZnO nanoparticles calcined at 200°C in air for 2 hours and is observed the diffraction peaks with the highest intensity. All diffraction peaks in the XRD patterns of as-prepared ZnO as well as calcined ZnO can be assigned to the hexagonal structure (JCPDS 36-1451).

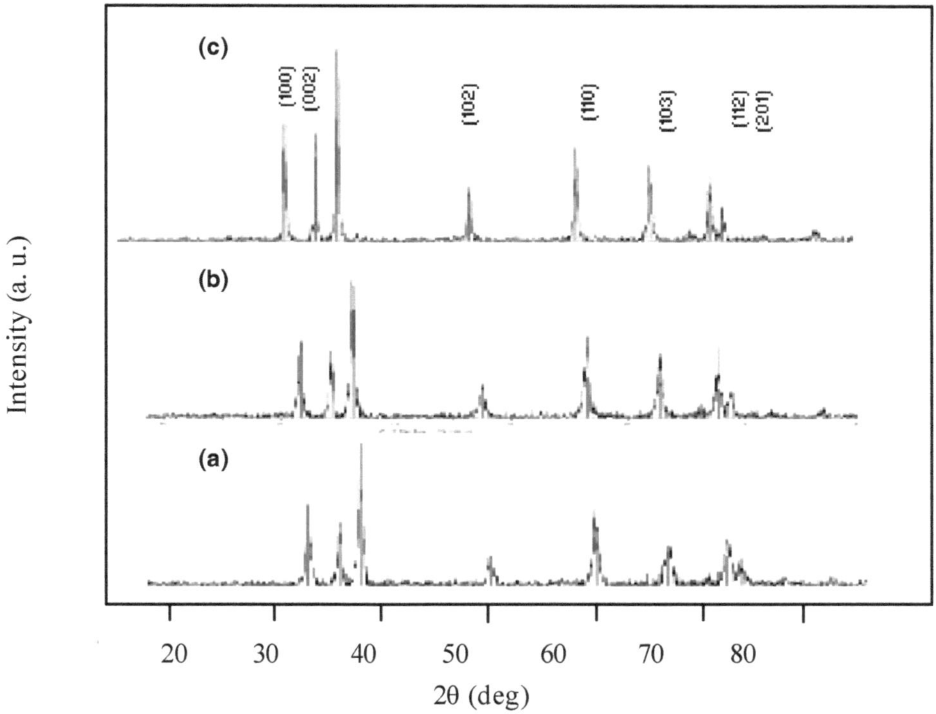

Figure 6.1: XRD Patterns of ZnO Nanoparticles Synthesized by Alkaline Precipitation at 60-70°C (a) as prepared for 1 hour reaction time, (b) as prepared for 2 hours reaction time and (c) calcined at 200°C for 2 hours.

No characteristic peaks were observed other than ZnO. It indicates that the high purity of the ZnO nanoparticles is obtained.

It was observed that the intensity of the diffraction peaks increased and became shaper with increasing in temperature, which implies that the crystalline structure tends to have more integrity (Aneesh, 2007). The average crystalline size of the products was estimated from XRD peaks by using the following Scherrer equation (Jenkins, 1996):

$$D = \frac{0.89 * \lambda}{\beta \cos\theta} \tag{1}$$

where λ is the wavelength of the incident X-ray beam; θ, the Bragg's diffraction angle; and β is the full width at the half-maximum (FWHM) of the ZnO (101) line in radians. As the reaction temperature increases, FWHM decreases. Thus, the size of ZnO nanoparticles increases as the temperature increases. This was due to the change of growth rate between the different crystallographic planes (Aneesh, 2007). By using this equation, the average sizes of the ZnO nanoparticles synthesized were found to be in the range of 14–20 nm depending on the crystal growth condition.

The SEM images of the ZnO nanoparticles prepared by alkaline precipitation technique were shown in Figures 6.2a to 6.2c. It was found that the morphology of the ZnO nanoparticles prepared is uniform and spherical shape. The zinc salt was soluble in water and the crystal growth was controlled by the diffusion of Zn^{2+} ions in the

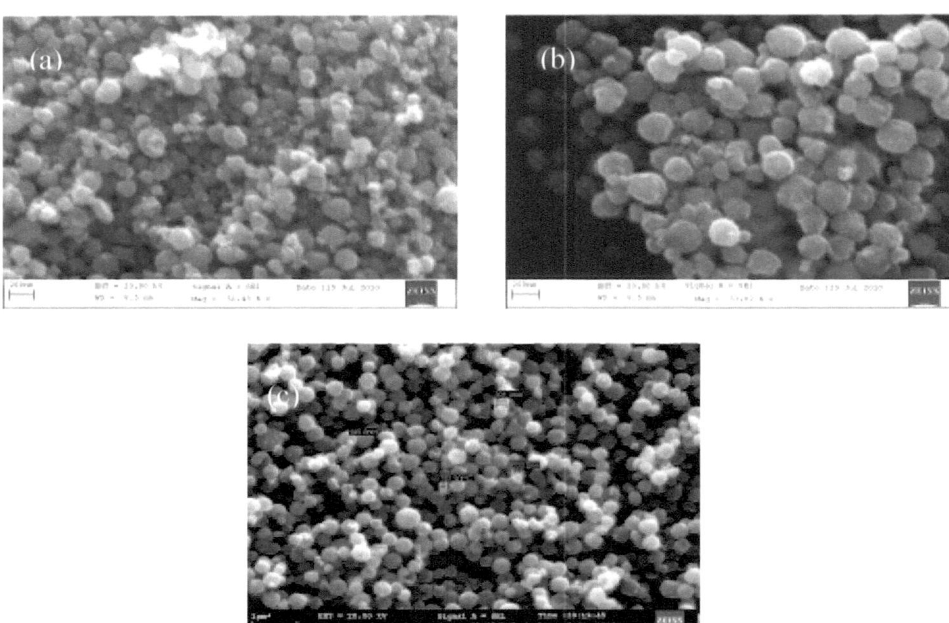

Figure 6.2: SEM Images of Zinc Oxide Nanoparticles Prepared (a) by Alkaline Precipitation for Reaction Time 1 h (b) for Reaction Time 2h, (c) After Calcination at 200°C for 2h

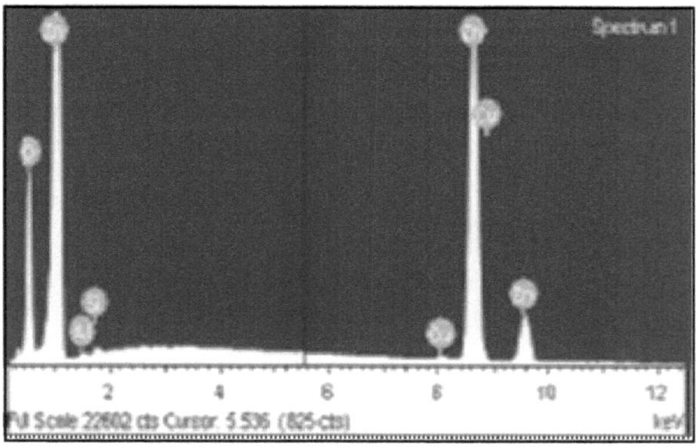

Element	*Weight per cent*
Zn	79.35
O	19.89
Si	0.24
Cu	0.41
Al	0.10
Total	100.00

Figure 6.3: EDX Spectrum and Elemental Analysis of ZnO Nanoparticles Prepared by Alkaline Precipitation

ethanol-rich layer. The TEA tended to make the particles more spherical. The size of the as-prepared particles by reaction times 1h was investigated less than 200 nm and the longer reaction time gave the larger particle sizes in the range from 150 nm to 250 nm. After calcinations at 200°C for 2h, the average particle size was found to be increased in the range 150 nm - 300 nm in diameter.

The elemental compositions of the ZnO nanoparticles prepared were determined by energy dispersive X-ray diffraction (EDX INCA Penta FETx3 OXFORD Instruments) as shown in Figure 6.3 indicating the high purity of ZnO nanoparticles.

Conclusions

The uniform and spherical shaped ZnO nanoparticles were synthesized using alkaline precipitation method. In this process, the TEA and alcohol were used as surface acting agent and as a dispersion medium respectively, which made the particles more spherical and uniform nanoparticles. The effect of the reaction time on the crystal structure and particle size of the nanoparticles was investigated. The XRD patterns and SEM images demonstrated that the crystal structure of the spherical ZnO nanoparticles was hexagonal and the particle sizes of the as-prepared ZnO nanoparticles were less than 200 nm. The particle size was found to be increased with the growth temperature as well as reaction time.

Acknowledgements

This research was supported by the Ministry of Science and Technology for the nanotechnology project. The authors are grateful to Mr. Khin Maung Win, XRD lab in MRDC, Dr. Aye Aye Toe, Mr. Myo Min Tun and Mr. Kyaw Naing Oo for taking keen interest in measuring of SEM/EDX and XRD in Analytical Lab, Pyin Oo Lwin.

References

Aneesh, P.M., Vanaja, K.A. and Jayaraj, M.K., 2007. Synthesis of ZnO nanoparticles by hydrothermal method. *Nanophotonic Matrials IV, Proc. of SPIE,* 6639: 66390J 1–9.

Hsieh, C.H., 2007. Spherical zinc oxide nano-particles from zinc acetate in the precipitation method. *J. of the Chinese Chemical Society*, 54: 31–34.

Hai, Y.X., Hao, W., Yong, C.Z. *et al.*, 2004. Hydrothermal synthesis of zinc oxide powders with controllable morphology. *Ceramics International*, 30: 93–97.

Husnu, E.U., Pritesh, H. and Nalin, R. *et al*, , 2008. Rapid synthesis of aligned zinc oxide nanowires. *Nanotechnology*, 19(255608): 1–5.

Jenkins, R. and Snyder, R.L., 1996. *Introduction to X-Ray Powder Diffractometry.* John Wiley and Sons, New York. pp. 55–60.

Nyffenegger, R.M., Craft, B. and Shaaban, M. *et al.*, 1998. A hybrid electrochemical/ chemical synthesis of zinc oxide nano-particles and optically intrinsic thin films. *Chem. Mater.*, 10: 1120–1129.

Satoshi, O., Tahereh, M., Mitsuo, U., Seiichi, T. and Tadafumi, A., 2004. Hydrothermal synthesis of fine zinc oxide nano-particles under supercritical conditions. In: *14th International Conference on the Properties of Water and Steam* in Kyoto, pp. 280–283.

Shah, M.A. and Al-Shahry, 2009. Zinc oxide nano-particles prepared by the reaction of zinc metal with ethanol. *JKAU Sci.*, 21: 61–67.

Yan, Z. and Yingxue, Z., 2008. Preparation of pure ZnO nano-particles by a simple solid-state rection method, Appl Phys. A: Material Science and Processing, 92, pp. 275–275.

Chapter 7

New and Advanced Manufacturing Technology: A Catalyst for Nigeria Technology and Economic Growth and Development

*Agoro, Olayiwola A.**

Department of Physical and Life Sciences,
Federal Ministry of Science and Technology,
Abuja, Nigeria

ABSTRACT

Nigeria quest for technology and economic growth and development dates back to pre-independence when the then government saw the need for the provision of basic infrastructures through the importation of technologies. The discovery of crude oil at post independence further propelled the necessity to source for technologies for the development of the Petroleum Sector among others. Over the years, the acquisition of foreign technologies has not met the required growth and development in the country due to non-adaptation and use of the technologies. With the entry into the market of new and advanced manufacturing technologies anchored by the National Agency for Science, Engineering and Infrastructures (NASENI), an agency of the Federal Ministry of Science and Technology, Nigeria is set to achieve the Vision 20:2020.

Keywords: *Nigeria Vision 20:2020, Infrastructural development, New and advanced manufacturing technologies.*

---

* E-mail: layiagoro@hotmail.com

Introduction

Nigeria like other developing countries is faced with the challenge of non acceptability or adaptability of technologies imported from developed countries. The quest for new and advanced technologies by nations including Nigeria was to further increase and improve the standard of living through the provision of basic and essential infrastructures for job and wealth creation, security and general well being of its citizens.

It is essential to state that the demand for development in the country will equally foster and implement the developmental plan of the country. Nigeria state had earmarked that in the next nine years to come, the country will strive to be among the 20 developed countries of the world with the development of all sectors among these being Science and Technology Sector. (NIP, Vision 20:2020).

Advanced Materials have always played an essential role in economic cycle, largely shaping every technological system. The development of advanced materials could lead to the design of completely new products such as medical inplants and computers among others. Exploitation of new and advanced materials technology will no doubt spring the needed investment in the country in the various sectors of the economy.

New and Advanced Materials Technology and its Sectoral Uses

The new and advanced materials technologies have been readily utilized in different sectors, among which are:

1. Advanced Manufacturing Technology
2. Virtual Manufacturing Laboratory
3. Computer Aided Design/Computer Aided Manufacturing
4. High Powered Computer Laboratory.
5. Rapid Prototyping
6. Nano-Technology
7. Capacity Development in Engineering Design and Development

Advanced Materials and their Applications

Advanced Materials as a potential technology in developing country such as Nigeria has the ability to change every life of its citizens. Potential areas where the technology had been adjudged useful to industrial applications in the developed countries will no doubt manifest itself in the country. These applications will in the future thus create job and wealth for the well being of the country. Materials such as ceramic, metals, polymers and biomaterials have all formed technology industrial advances. This technology will have enormous impact in manufacturing, electronics, and IT.

New and Advanced Materials Applications Areas

New and Advanced Materials application areas are diverse. The following are widely utilized areas:

- ✰ Antennae that bends towards a signal;
- ✰ Self repairing house;
- ✰ Cans which can be crushed and regain their original shape under heat so that they can be recycled;
- ✰ Smart materials in clothes we wear could monitor ones health or stress levels etc;
- ✰ Vehicles contains many advanced materials including plastics and alloys such as ceramic used in reducing wear, oil with additives maintain lubrication, polymer used in interiors etc;
- ✰ Bicycle parts such as spokes, tyres, seat and frames are all made up of advanced materials;
- ✰ Medical applications of advanced materials in dentistry or make better contact lenses;
- ✰ Sunscreens uses nanosized particles to absorb and reflect ultraviolet rays as applicable in body lotions.
- ✰ Self Cleaning Windows that are coated with materials. In this case, when the sun shines on these windows chemical reaction is triggered which breaks down the dirts, and gets it washed down, this could be attributed to nanoscale.
- ✰ Stain Repellent Fabrics: These are made by immersing rolls of woven cotton fabric in liquids that contains millions of nanotech fibres. Such cottons are dried in Oven thus binding the tiny fibres to the cotton threads. The cotton thus becomes impermeable to liquid.
- ✰ Bouncy Tennis Balls are such that are coated in nanosized material. In this tiny particles which trap air molecules makes the balls to bounce extra.

New and Advanced Technology and its Effects on the Economy

The Nigeria government as a way of industrializing the country recognizes the critical role of applied technology, research and innovation in aiding the expansion of its economy. It should be noted that industrialized countries embarked on different avenues to getting industrialized for example, Brazil took the way of Western Europe and United States by making Science and Technology part of their general education and hi-technology part of modern professions. South Korea on their own focused on technology with focus on its universities and R&D institutions,

In all the countries mentioned above, investments in technology were made in industrial firms coupled with isolated government agencies. The countries had recognized that industrialization was key to economic growth and development with the backbone of industrialization being Research, Technology and Innovation.

Nigeria too, is gradually becoming aware that industralisation means driving the Science, Technology and Innovation in the right direction through putting in place measures aimed at expanding capacity and capability (Vision, 20:2020)

Nigeria Path to Economic Growth through Advanced Materials

Having recognized that the only path to industralisation was through the development of the science and technology, Nigeria as a country is focusing on the development of advanced materials in generating the needed employment and as a way of industrializing the nation. The nation had focused on the development of these projects:

1. Development of process technologies for the production of industrial chemicals from petrochemicals;
2. Construction of Printed Circuit Board for computer motherboards and mobile phones;
3. Establishment of advanced multipurpose centres in 6 zones of the country.

Also, the government through the Science, Technology and Innovation had recognized priority projects and programmes cutting across all science sectors. Priority is scheduled to be given new and advanced materials.

The priority programmes and projects and their numbers are as stated below:

1. Improving National R&D Activities (9)
2. Energy Research and Development (11)
3. Advanced Materials and Manufacturing (8)
4. Geo Science Research (4)
5. Small and Medium Enterprises (12)
6. Human Resource Development (18)
7. Standardization and Quality Assurance (10)
8. Bio and Nano-technology (8)
9. STI Infrastructure (26)
10. Health (12)
11. Agriculture (6)
12. Logistics (7)

Challenges Hindering Industrilization of Advanced Materials in Nigeria

- The required number of Researchers are inadequate for the industrialization in the country;
- The required funding to kick start the use of nanotechnology is inadequate;
- There is uncoordinated and indiscriminate application of resources for its application;
- Absence of linkage between Government, Industry and Research Institutions;
- Inadequate and low level of capacity.

Institutional Framework Involved in Driving New and Advanced Technology in Nigeria

In Nigeria, the National Agency for Science and Engineering Infrastructure (NASENI) is at the moment driving the promotion of research and development and the application of New and Advanced Materials Technology in the country.

The Agency under the purview of the Federal Ministry of Science and Technology was established in 1992 with the mandate of making available the primary and intermediate products required for machine and equipment design, fabrication and even mass production to achieve a sustainable industrialization in the country. As a way of achieving the required industrialization in the country, the Agency was mandated to establish Institutes in R&D in the various fields of Science and Engineering with target to empower the small and medium scale industries through impartation of technologies, engineering principles and practices for the production of equipment.

Development Institues/Centres under NASENI

1. Science Equipment Development Institute (SEDI), Enugu
2. Science Equipment Development Institute (SEDI), Minna
3. Electronic Development Institute (ELDI), Akwa
4. Hydraulic Equipment Development Institute (HEDI), Akure
5. Engineering Materials Development Institute (EMDI), Akure
6. National Engineering Design Development Institute (NEDDI), Nnewi
7. Power Equipment and Electrical Machinery Development Institute (PEEMADI), Okene
8. Prototyping Engineering Development Institute (PEDI), Ilesa
9. Advanced Manufacturing Technology Project, Jigawa

Operations of Natoinal Agency for Science and Engineering Infrastructure (NASENI)

The Agency, its Development Institutes and Centre are mandated with the following operations in order to achieve the required industralisation in the country.

- ☆ Engineering materials (notably irons, steels, non-ferrous metal and alloys, plastics, glass, ceramics, polymer electronics and nanotechnology); Industrial and analytic chemical materials including industrial gases;
- ☆ Scientific equipment and components for education, research and industry; including measuring instruments, electronic components, communication equipment and computers;
- ☆ Engineering accessories (mechanical, hydraulic, pneumatic, electrical and electronic);
- ☆ Engineering Design of components parts, plants and processes.

- ☆ Power equipment (generation, transmission, distribution, prime movers); and
- ☆ Mechanical Engineering tools (power tools, hand tools, cutting tools and machine tools).

Nanotechnology Research and Development Capacity Building Programmes

1. Linkage with AMRS, Training of 15 Universities Lectures in American Universities (Princeton, Maryland), Sheda Science and Technology Complex (SHETSCO), University of Ibadan (UI), Abubakar Tafawa Balewa University (ATBU) and Federal University of Technology, Owerri (FUTO).
2. Nano-particles production for energy (Solar cells and light emitting diodes), Structural materials and Biology(Medicine, agriculture and environment)
3. National Training for Universities in Nano-technology
4. Research in residence programe Nano-technology(6 months residency)
5. Sabbatical in NASENI

Way Forward and Future Strategies

The advent of new and advanced technologies cum materials in the different fields will no doubt expand the scope of industrialization in the country. The Federal Government of Nigeria will be able to create the needed employment and equally

Figure 7.1: High Power Computing Laboratory at the NASENI Headquaters, Abuja

Figure 7.2: CNC Lathe Machine at the NASENI Headquarter Used for Cutting/ Grinding Metals Used in Industries

Figure 7.3: An Engineer Inspecting an Electrical Discharge Machine (EDM) at the NASENI Headquarter Used for Cutting Pre-hardened Steel without need for Heat Treatment for Research and Development

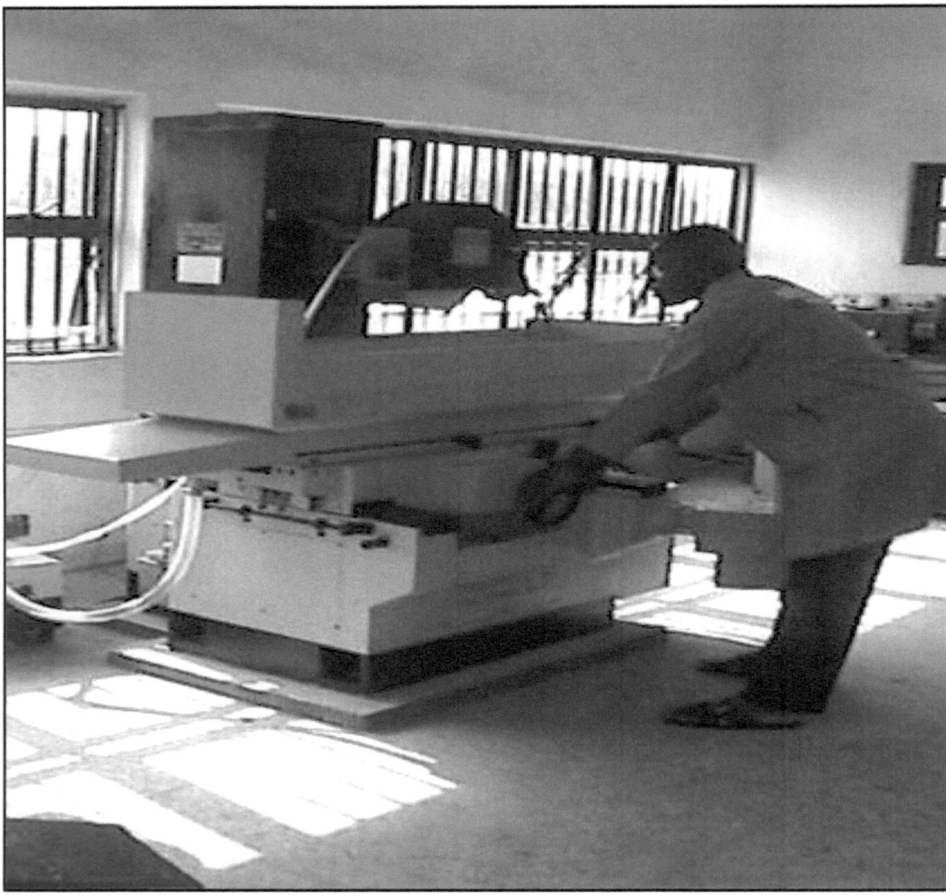

Figure 7.4: An Engineer at the NASENI Headquarter Working on Surface Grinding Machine Used for Cutting and Grinding Metals

create the much needed wealth if advanced materials are properly harnessed. Nigeria's aspiration to be among the top 20 leading economies can only be realized when this is done. The Nigeria's quest towards achieving that feat will be realized once resources are utilized towards this direction.

References

1. Adewoye, O.O., 2011. *First Ministerial Scientific Briefing*, Transcorp Hilton Hotel. 26th February.
2. Cohendet, P., Ledoux, M.-J. and Zuscovitch, E., 1987. Les matriaux nouveaux: dynamique conomique et stratgie europenne. Paris: *Economica.*
3. Nigeria Vision 20: 2020, 2010. Economic Transformation Blueprint. 1 National Implementation Plan 2010–2013.

Chapter 8

Bagasse (Sugar Cane Waste): As Low Cost Reinforcement for Advanced Metal Matrix Composites

V.S. Aigbodion and S.B. Hassan*

Department of Metallurgical and Materials Engineering, Ahmadu Bello University, Samaru, Zaria, Nigeria

ABSTRACT

Despite the potential benefits of lighter weight and durability resulting from corrosion resistance, advanced composites are not recognized as a material of choice in the near term for automotive applications. Significant changes on a broad spectrum would be required to make advanced composites attractive for widespread commercial use in cars and trucks. The principal barrier is the high cost of the raw and converted materials when compared to existing options, and the perception that even with volume production in yet-to-be-perfected processing methods, the costs would still be too high. In attempts to overcome this limitation, several research and development (R&D) were focused on the reinforcement of aluminum-based MMCs using low cost industrial waste by-products as the reinforcement. This could reduce the cost and the weight of energy intensive metals for potential applications in engineering components for a new generation of vehicles. This paper summarizes attempts of incorporating bagasse into aluminum castings to decrease the energy content, material content, cost, and weight of selected industrial components, while also improving selected properties. The aluminum matrix composites reinforced with bagasse ash was produced by double stir-casting. 2-10 weight percent bagasse ash particles were added. The microstructure and phases of the composites produced were examined by SEM/EDS and XRD methods.

* *Corresponding Author:* E-mail: aigbodionv@yahoo.com

Properties such as density, hardness, wear, tensile compression strength of the were evaluated. These composites can be used in area of wear resistance i.e in the production of pistons, cylinder liners, brake discs and bushing etc. since aluminum composites used in production of these components has similar properties to that of Al-Cu-Mg/BAp composites.

Keywords: *Mechanical properties, Micro-structure and Wear.*

Introduction

The high cost of current metal matrix composites(MMCs), interfacial reaction and high density of most common ceramic reinforcement compared to aluminum alloys has inhibited production on a large industrial scale, for example in the automotive industry (Aigbodion, 2010). In attempts to overcome this limitation, several Research and Development (R&D) programme were focused on the reinforcement of aluminum-based MMCs using low cost industrial waste byproducts as the reinforcement particulate. This could reduce the cost and the weight of energy intensive metals for potential applications in engineering components for a new generation of vehicles. However, the future commercialization of these materials in automotive industry will be also influenced by the ability to recycle the scrap and the final components when its useful life is over. Thus, researchers have worked out separately to reinforce SiC, Al_2O_3 (*i.e.* carbides, Nitrides and oxides) TiB_2, Boron and Graphite into the Aluminium matrix to achieve different properties and they are expensive and of high density (Figure 8.1) (Bienia *et al.*, 2003). Because of this researcher are looking for low cost reinforcer, that will be the cost of composites to a low level.

The ever-increasing demand for low cost reinforcement stimulated the interest towards production and utilization of using bye-products from industry as reinforcement since they are readily available or are naturally renewable at affordable cost. Aigbodion(2007) have used Kankara clay (alumino-silicate) in reinforcing Aluminium-Silicon (Al-Si) alloy, Bienia *et al.* (2003) used fly ash in reinforcement of aluminium matrix, Naresh(2006), worked on the development and characterization of metal matrix composite using red mud an industrial waste for wear resistant

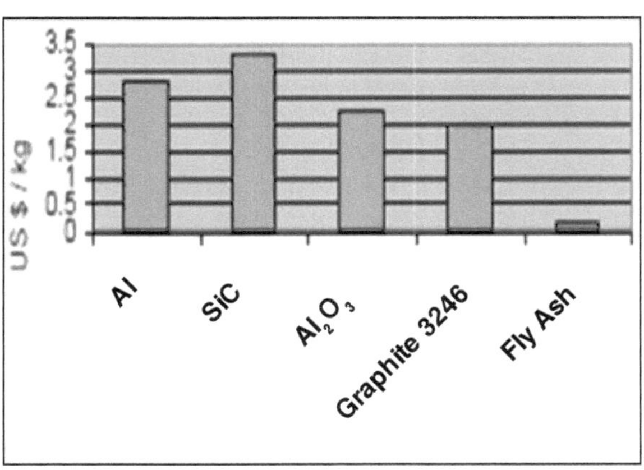

Figure 8.1: Prices of Materials Commonly Used in making MMCs

applications, they all reported good dispersion and recovery of the particles in the composites castings. It is in the light of the foregoing researches that investigation into the possibility of using bagasse ash in metal matrix particulate composite for engineering applications was motivated.

Bagasse

Bagasse is the residue fiber remaining when surgar cane is pressed to extract the sugar. Some bagasse is burned to supply heat to the sugar refining operation. Some is returned to the fields, and some finds it way into various board products. Bagasse is composed of fiber and pith. The fiber is thick walled and relatively long (1-4 mm). Fibers are obtained mostly from the rind, but there are fibrovascular bundles dispersed throughout the interior of the stalk as well (Aigbodion, 2010). Until recently, the remaining 90 per cent (empty fruit bunches, fibers, fronds, trunks) was discarded as waste, and either burned in the open air or left to settle in waste ponds. This way, the Sugar-cane processing industry's waste contributes significantly to CO_2 and methane emissions.

Bagasse is available wherever sugarcane is grown. As such, almost no harvesting problems exist, and large volumes are available at sugar mills. In northern climates, the cane harvest usually lasts about 2.5 months. In warm climates, bagasse may be available for as long as 10 months out of the year. During this time, bagasse supply is relatively constant: the remainder of the year, it must be stored. To reduce the sugar content and increase storage life, bagasse is usually depithed before storage (Aigbodion, 2010).

Materials and Methods

Materials and Equipment

The bagasse used in this work was obtained from Zongo area of Zaria in Kaduna-State Nigeria (Figure 8.2). The bagasse was then carbonized at 1200oC for 5hours to obtain a black color ash which is the bagasse ash and sieve to average particle size of 63µ m (Figure 8.3) and the X-ray diffraction analysis (XRD) of the bagasse ash was determined. The X-ray diffractogram of the bagasse ash is presented in Figure 8.4. Magnesium power, high purity aluminum and copper electrical wire are obtained from Northern Cable Company NOCACO (Kaduna), Moulding boxes, Silica sand and Bentonite are obtained from National Metallurgical Development Centre, Jos, Nigeria.

The equipment used in this study included: Pyrometer, mechanical stirrer, crucible, electrical resistance furnace, Rockwell hardness tester, Charpy impact machine, Tinus Olsen tensile machine, Scanning electron microscope (SEM) with energy dispersive X-ray spectroscopy (EDS) and pin on disc machine.

Procedure

Composite used in this study was Al-Cu-Mg–Bagasse Ash particles composite containing 2-10 wt per cent bagasse ash particles at interval of 2wt per cent. The samples were produced using the double stir-casting method (Aigbodion, 2010) by

Figure 8.2: Photomacrograph of the Bagasse

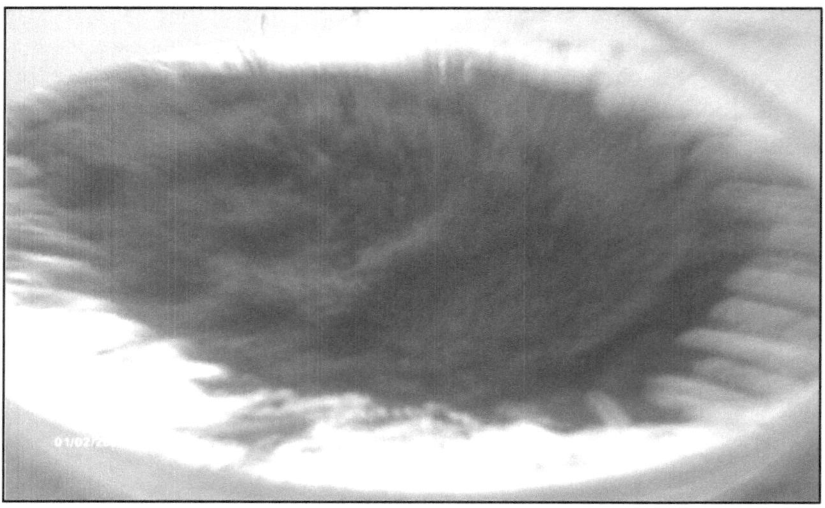

Figure 8.3: Photomacrograph of the Bagasse after Carbonization

keeping the percentage of copper and magnesium constant (3.7 per cent Cu and 1.4 per cent Mg) according to the recommended standard to produce alloy of type A2009. A control sample without the bagasse ash particles was also produced in this method. A preheated sand mould with diameter 18 mm and 300mm length was used to produce cast bars. After casting, the samples were machined into tensile, impact and hardness test samples for the purpose of determining the mechanical properties. The various diffraction peaks, inter-planar distance and phases in the alloy and composites with 8wt per cent bagasse ash particles were determined using a Bruker D8 θ-θ X-ray diffractometer.

The microstructure and the chemical compositions of the phases present in the test samples were studied using a JEOL JSM 5900LV Scanning Electron Microscope equipped with an Oxford INCATM Energy Dispersive Spectroscopy (EDS) system.

The basic method of determining the density of a sample by measuring the mass and volume of the sample was used. The density of the sample was then estimated from the formula given below:

$$\text{Density} = \text{Mass}/\text{Volume} \tag{1}$$

The hardness values of the samples were determine (ASTM E18-79) using the Rockwell hardness tester on "B" scale (Frank Well test Rockwell Hardness Tester, model 38506) with 1.56mm steel ball indenter, minor load of 10kg, and major load of 100kg and hardness value of 101.2HRB as the standard block.

The tensile properties of the as-cast composites sample were conducted on Tinus-Olsen tensile testing machine with a strain rate of 2x10-3S-1. The test pieces were machine to the standard shape and dimensions as specified by the American Society for testing and Materials.

The impact test of the as cast composites sample was conducted using a fully instrumented Avery Denison test machine. Charpy impact tests were conducted on notched samples. Standard square impact test sample measuring 75 x 10 x 10 mm with notch depth of 2 mm and a notch tip radius of 0. 02 mm at angle of 45o was used.

A pin-on-disc test apparatus was used to investigate the dry sliding wear characteristics of the unreinforced and reinforced Al-Cu-Mg/BAp alloy as per ASTM G99-95 standards. Wear specimen 20 mm in diameter and 40 mm high was cut from as-cast specimens, machined, and then polished metallographically. Wear tests were conducted with loads ranging from 5 to 20 N and sliding speeds of 1.26 to 5.02m/s for a constant sliding distance of 5000m. The frictional traction experienced by the pin during sliding was measured continuously by a PC-based data-logging system. After running through a fixed sliding distance, the specimens were removed, cleaned with acetone, dried, and weighed to determine the weight loss due to wear. The differences in weight measured before and after tests give the wear of the specimen. The formula used to convert the weight loss into wear rate is given in the equation below:

$$\text{Wear rate} = \frac{\Delta W}{S} \tag{2}$$

where, ΔW is the weight difference of the sample before and after the test in kg, S is total sliding distance in m.

Results and Discussion

Figure 8.4, shows the X-ray diffractogram of the bagasse ash. From the figure it was observed that, the diffraction peaks of the bagasse ash were 20.68°, 26.53°, 35.41° and 40.00° and their inter-planar distances were 4.29Å, 3.36Å, 2.54Å and 2.26Å. Phases at these peaks as Quartz (SiO_2), Cliftonite(C), Moissanite (Si C) and Titanium Oxide (Ti 6O). The mineralogical analysis of bagasse ash carried out by XRD reveals

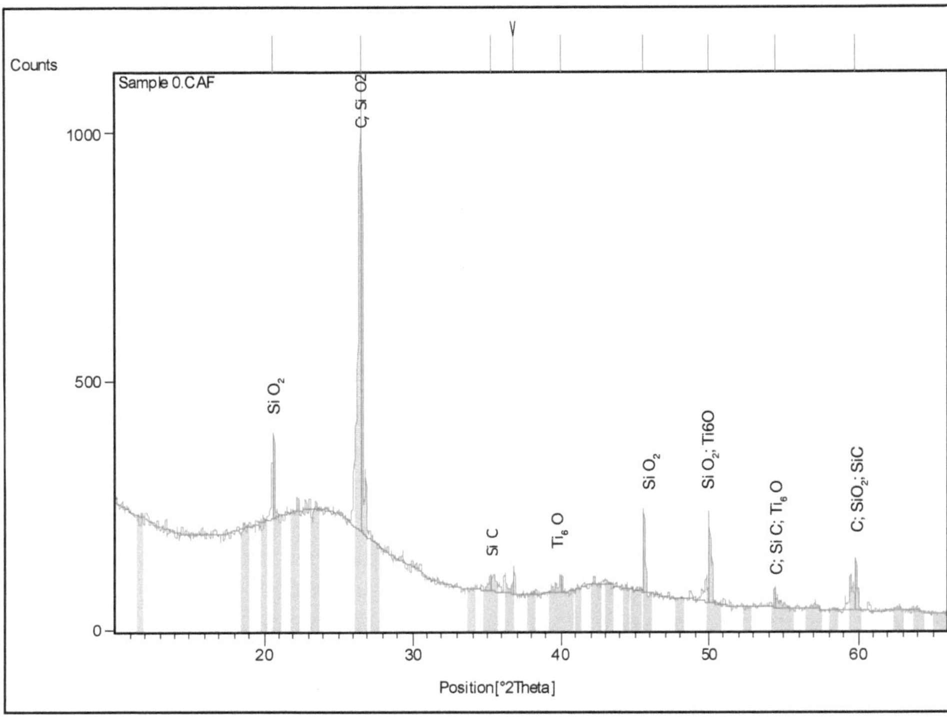

Figure 8.4: XRD Pattern of the Bagasse Ash

C, SiO_2, SiC and Ti6O as the major constituents and the minor ones are Fe_2O_3, Na_2O, CaO and MnO (Table 8.1). From Table 8.1 it was observed that C liftonite(C) has the higher score percentage of 31 per cent.

Table 8.1: Identified Patterns List of the XRD of Bagasse Ash

Visible	*Ref. Code*	*Score*	*Compound Name*	*Displacement [°2Th.]*	*Scale Factor*	*Chemical Formula*
*	00-041-1487	31	Cliftonite	0.000	0.657	C
*	01-085-0798	30	Quartz	0.000	0.084	SiO2
*	01-075-1541	19	Moissanite 6\ITH\RG	0.000	0.044	Si C
*	01-072-1807	18	TitaniumOxide	0.000	0.045	Ti6O

Macrostructural studies revealed a uniform distribution of bagasse ash particles. The distribution of bagasse ash particles is influenced by good wettability of the bagasse ash by the molten metal and good interfacial bonding between bagasse ash particles and matrix material (Figure 8.5).

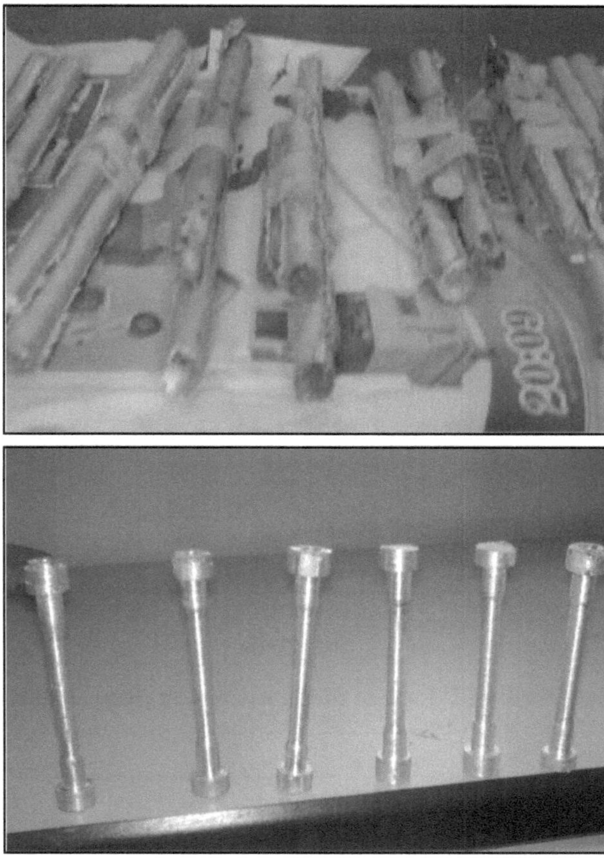

Figure 8.5: The Alloy and Composites after Casting and Machining

The good wettability obtained in this study was a result of the use of a covering fluxes combined with double stirring method during casting which help to avoid gravity segregation. This prevented the composite slurries from oxidation and breakdown of the gas layer. The XRD pattern of the unreinforced Al–Cu-Mg alloy (A2009) and Al–Cu-Mg alloy reinforced with 8wt per cent bagasse ash composite manufactured by double stir casting method are shown in Figures 8.6–8.7.

From the Figures 8.6–8.7 it was observed that, the major diffraction peaks of the unreinforced alloy are 37.6881°, 43.8769° and 64.4819° and their inter-planar distance, 2.38685Å, 2.06348Å and 1.44391Å, phases at these peaks are Alumimum (α-Al), Aluminum Copper p has e(Cu_3Al_2) and Aluminum Copper Magnesium ph as e(Al6CuMg4). The composite shows three major principal diffraction intensity of $2\theta=37.7722°$, 44.0750° and 64.3863° and their inter-planar distances are 2.38173Å, 2.05466Å and 1.44582Å, phases at these peaks are silicon carbide (SiC), Aluminum Titanium Carbide ph as e ($AlTi_3C$) and Aluminum Copper p has e(Cu_3Al_2). In these diffractograms, one can evidently deduce the crystalline phases of the master alloy

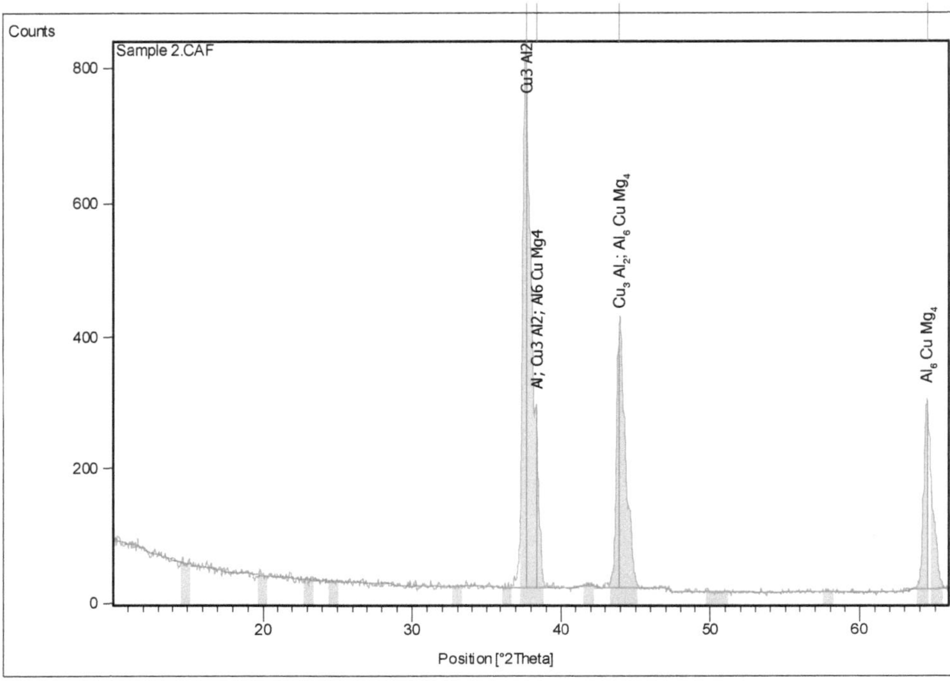

Figure 8.6: XRD Pattern of the Aluminum Alloy (A2009)

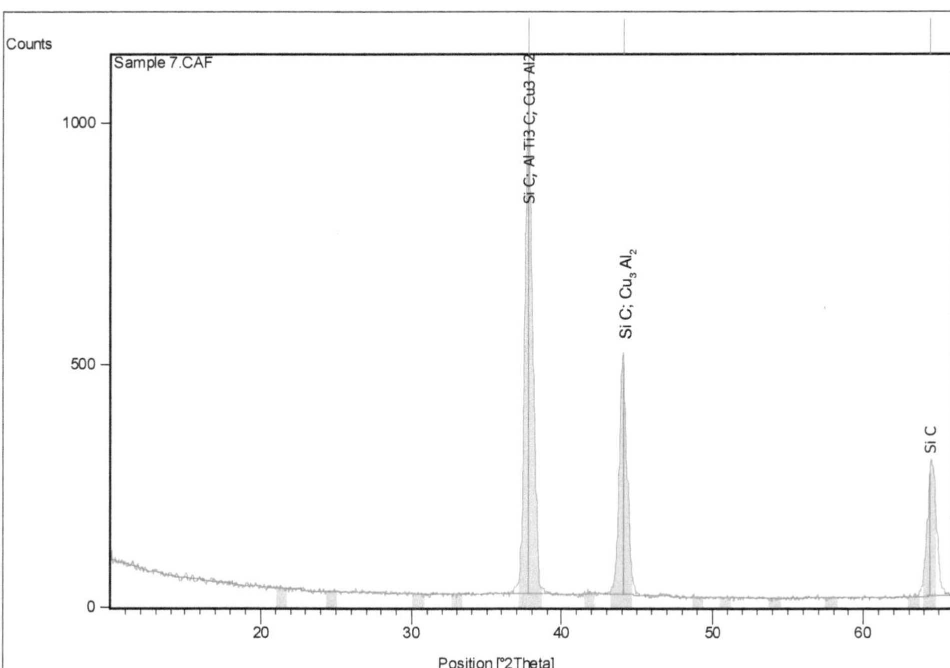

Figure 8.7: XRD Pattern of the Aluminum Alloy (A2009) Reinforced with 8wt per cent Bagasse Ash

from that of the composite material. The X-ray patterns show the α-aluminum, copper and magnesium presence on the matrix alloy Al–Cu- Mg and evident of the bagasse ash particles in the composite.

The microstructure of the unreinforced Al–Cu-Mg alloy is shown in Figure 8.6. The structure reveals the eutectic phase containing Cu_3Al_2, Al_6CuMg_4 in α-aluminum matrix. In the Al-Cu-Mg alloy, Cu and Mg are present both in solid solution as well in precipitated form as Cu_3Al_2, and Al_6CuMg_4 phase both in the grain and along the grain boundaries. Figures 8.8–8.13, show the microstructure of the reinforced alloy with bagasse ash particle additions. The microstructure reveals that there are reasonably uniform distributions of bagasse ash particles in the metal matrix. The ceramic phase is shown as dark phase, while the metal phase is white (Figures 8.9–8.13). However, there is an agglomeration and segregation of bagasse ash particles in the microscale with alloy reinforced with 10wt per cent bagaase ash particle (Figure 8.13).

The unreinforced alloy reveals the presence of α - A l, Cu and Mg as evident from the EDS spectra (Figure 8.4). From the EDS analysis of the composites material(see Figures 5-10) there is an indication of some possible chemical reaction between aluminum melt and bagasse ash particles which led to the release of Si, Fe, Na, O and K etc. in the composites, since the bagasse ash particles consist of C, SiO_2, SiC and Ti_6O as the major constituents, while the minor ones are Fe_2O_3, Na_2O, CaO and MnO (Figure 8.1) these constituents react with Al and Cu present in the molten matrix alloy depending on the kinetics of reactions. Some of the possible interfacial reactions are:

$$4Al(l) + 2C(s) + Ti_6O(s) + Cu(l) = Al_2CuO(s) + 2AlTi_3C(s)$$

$$(\Delta G = -51.2\ kJ/kg \text{ at } 1000\ K)$$

It should be noted that the presence of the bagasse ash particles result in a much smaller grain size in the cast composites compared to the Al-Cu-Mg alloy grain size. Similar observations have been made by Rajan *et al.* (2007) on commercially pure Al alloy based fly ash composites.

The density decreased with increasing percentage additions of bagasse ash particles (Figure 8.14). The density of bagasse ash was 1.95g/cm^3 and the overall density of bagasse ash composites was decreased with wt per cent additions of bagasse ash. This shows that composites with light weight can be made with this ash.

The hardness values of the Al-Cu-Mg/BAp of the composite increased with an increasing percentage of bagasse ash particle additions. This is noteworthy that the hard value of bagasse ash is 75.5HB and the presence of the hard ceramic phase in the ductile matrix has resulted into the increase in the hardness of the composite (Figure 8.15).

The 10wt per cent bagasse ash addition to the alloy yielded the highest hardness value. As far as hardening behavior of the composites is concerned, particle addition in the matrix alloy increases the strain energy in the periphery of the particles in the matrix and these tendencies may be due to the formation of the dislocation at the

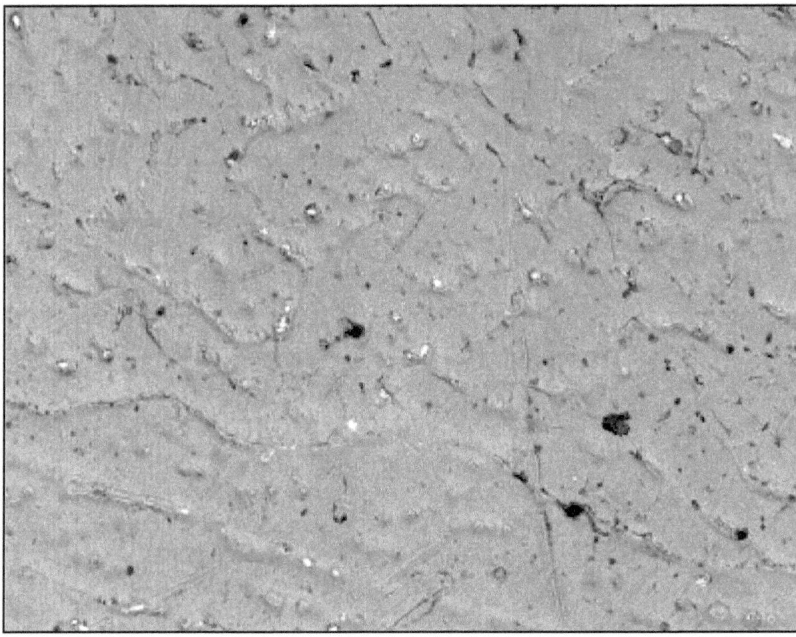

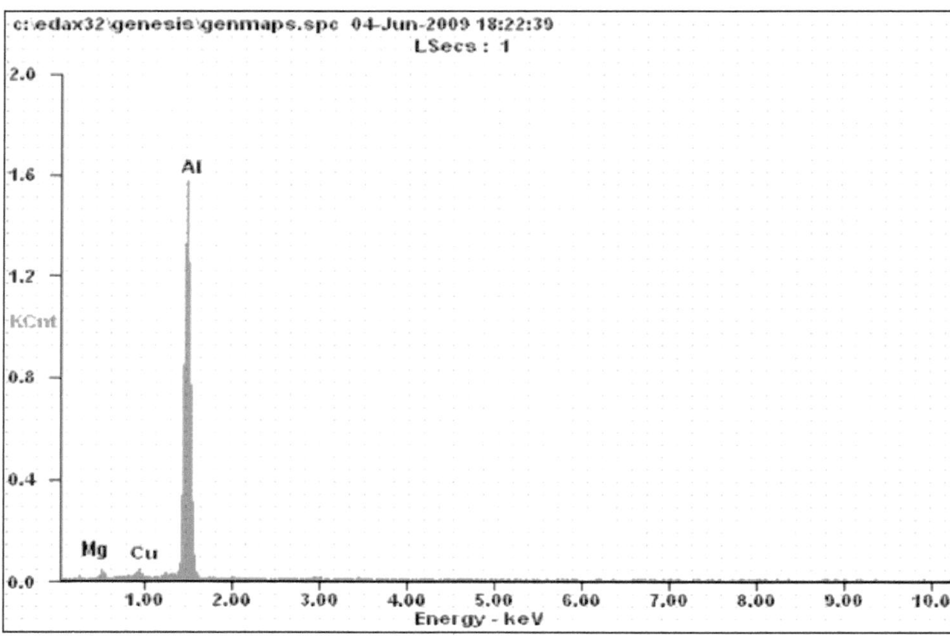

Cu_3Al_2, and Al_6CuMg_4 in α-Al

Figure 8.8: SEM/EDS of Microstructure of the Uinreinforced Al-Cu-Mg Alloy

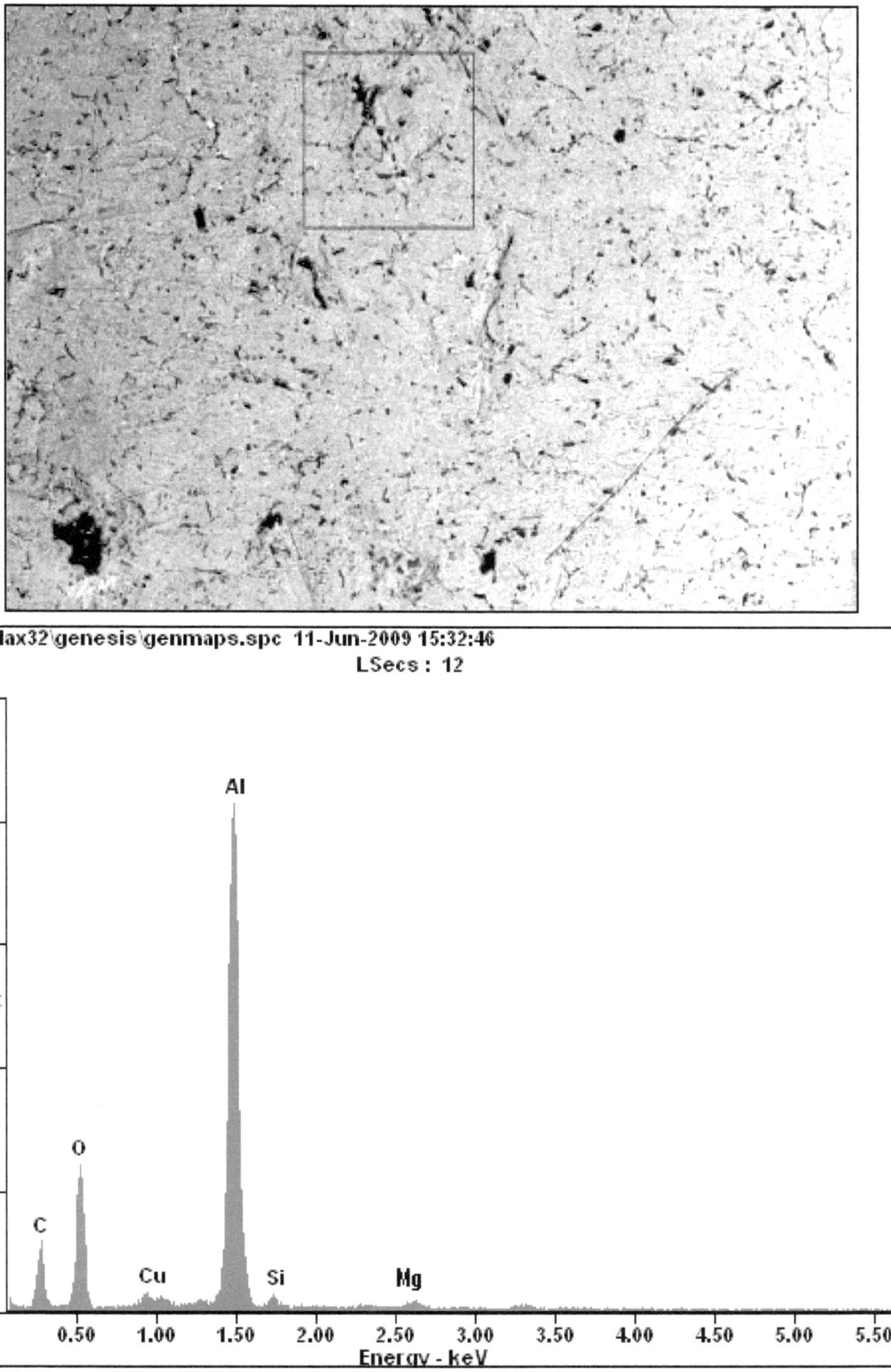

BAp

Figure 8.9: SEM/EDS of Microstructure of the Reinforced Al-Cu-Mg Alloy with 2wt per cent Bagasse Ash

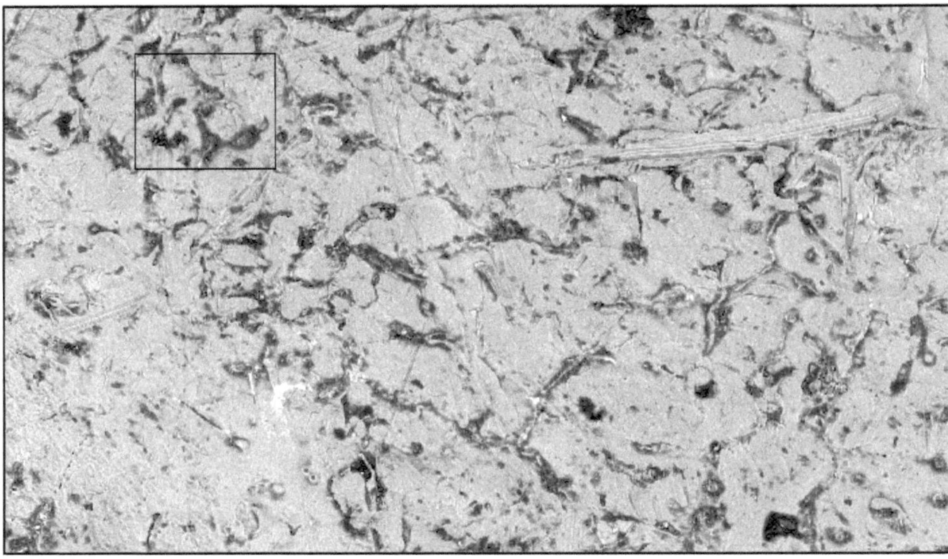

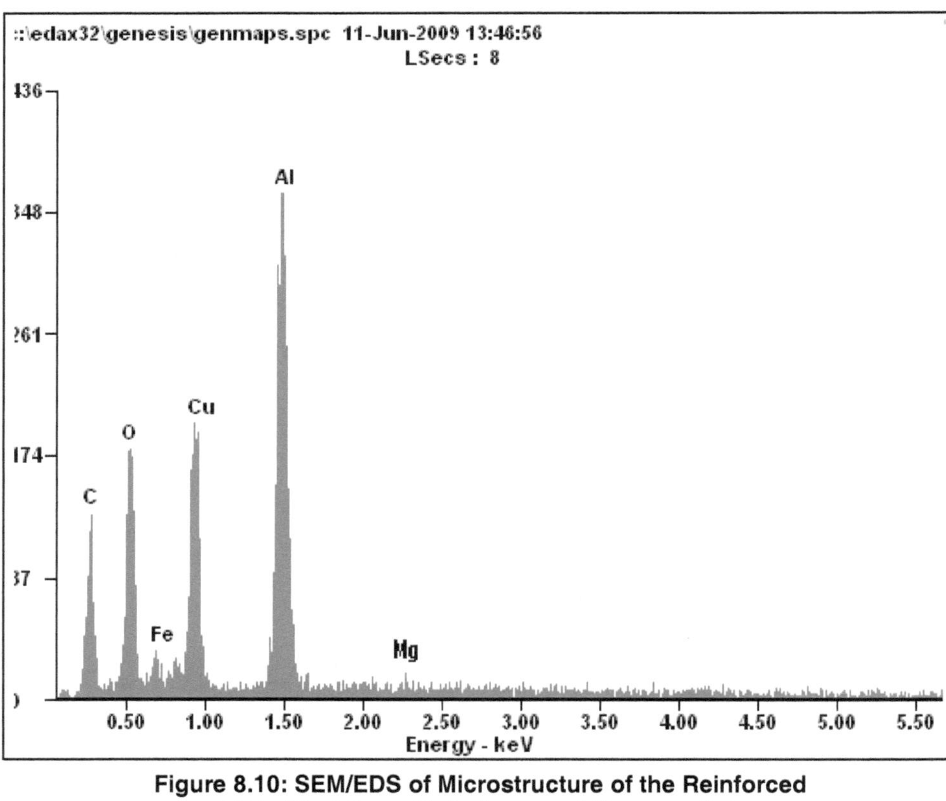

Figure 8.10: SEM/EDS of Microstructure of the Reinforced Al-Cu-Mg Alloy with 4wt per cent Bagasse Ash

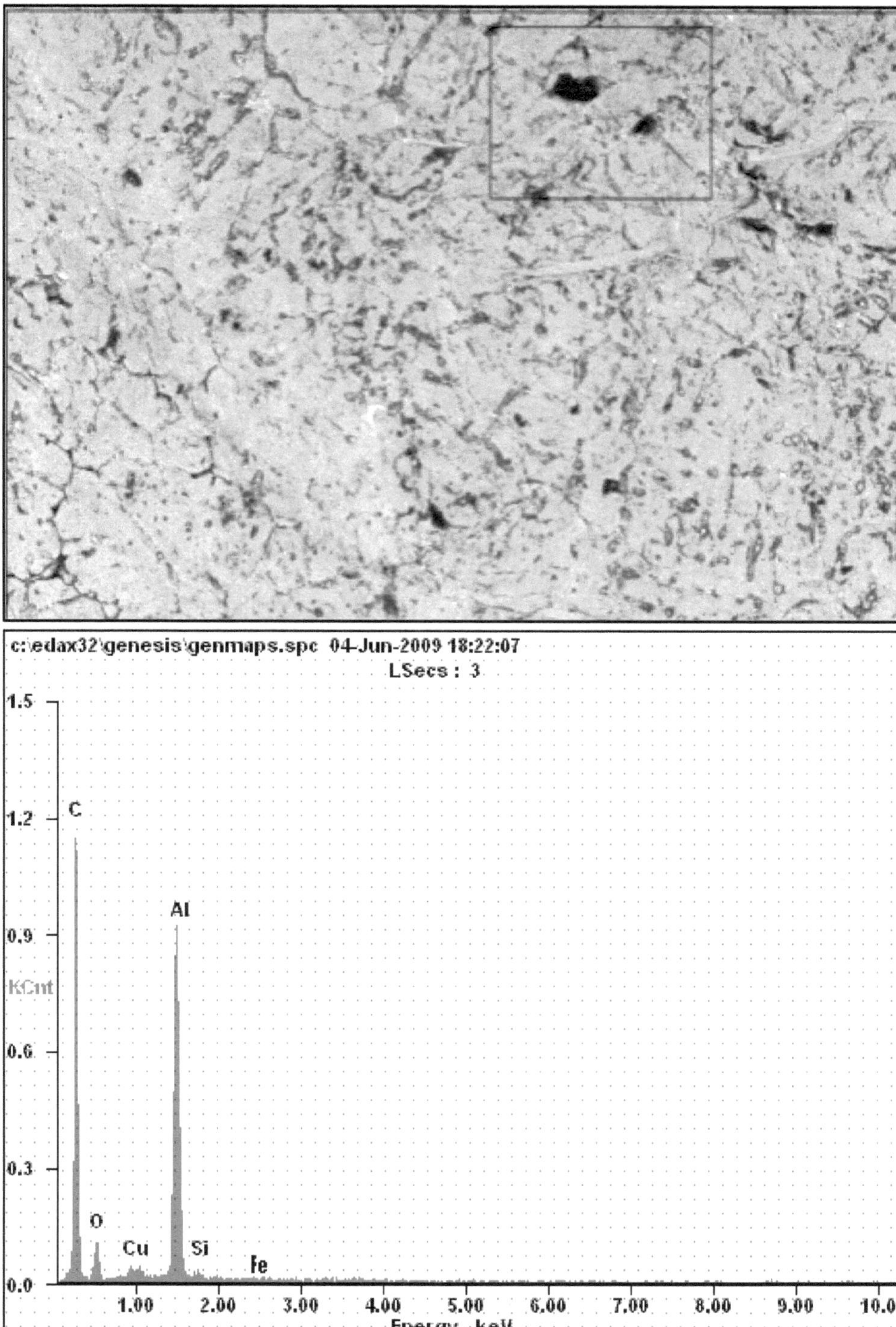

Figure 8.11: SEM/EDS of Microstructure of the Reinforced Al-Cu-Mg Alloy with 6wt per cent Bagasse Ash

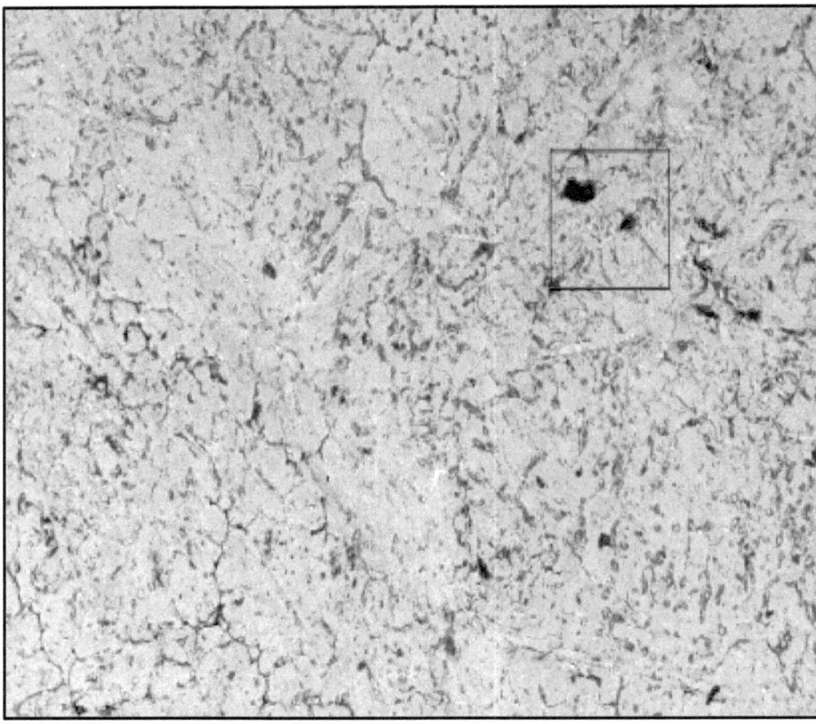

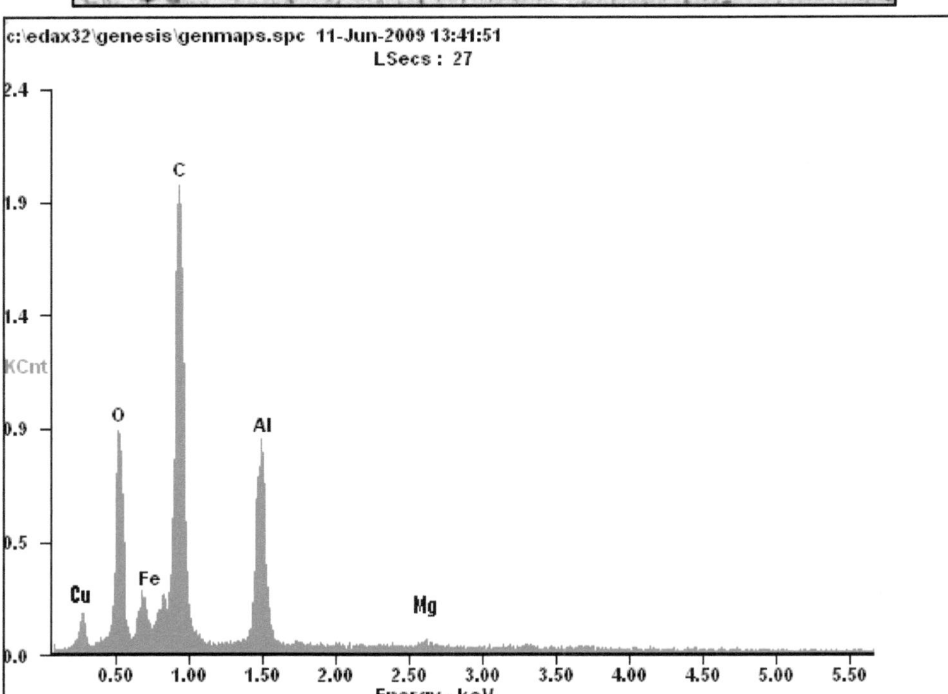

Figure 8.12: SEM/EDS of Microstructure of the Reinforced Al-Cu-Mg Alloy with 8wt per cent Bagasse Ash

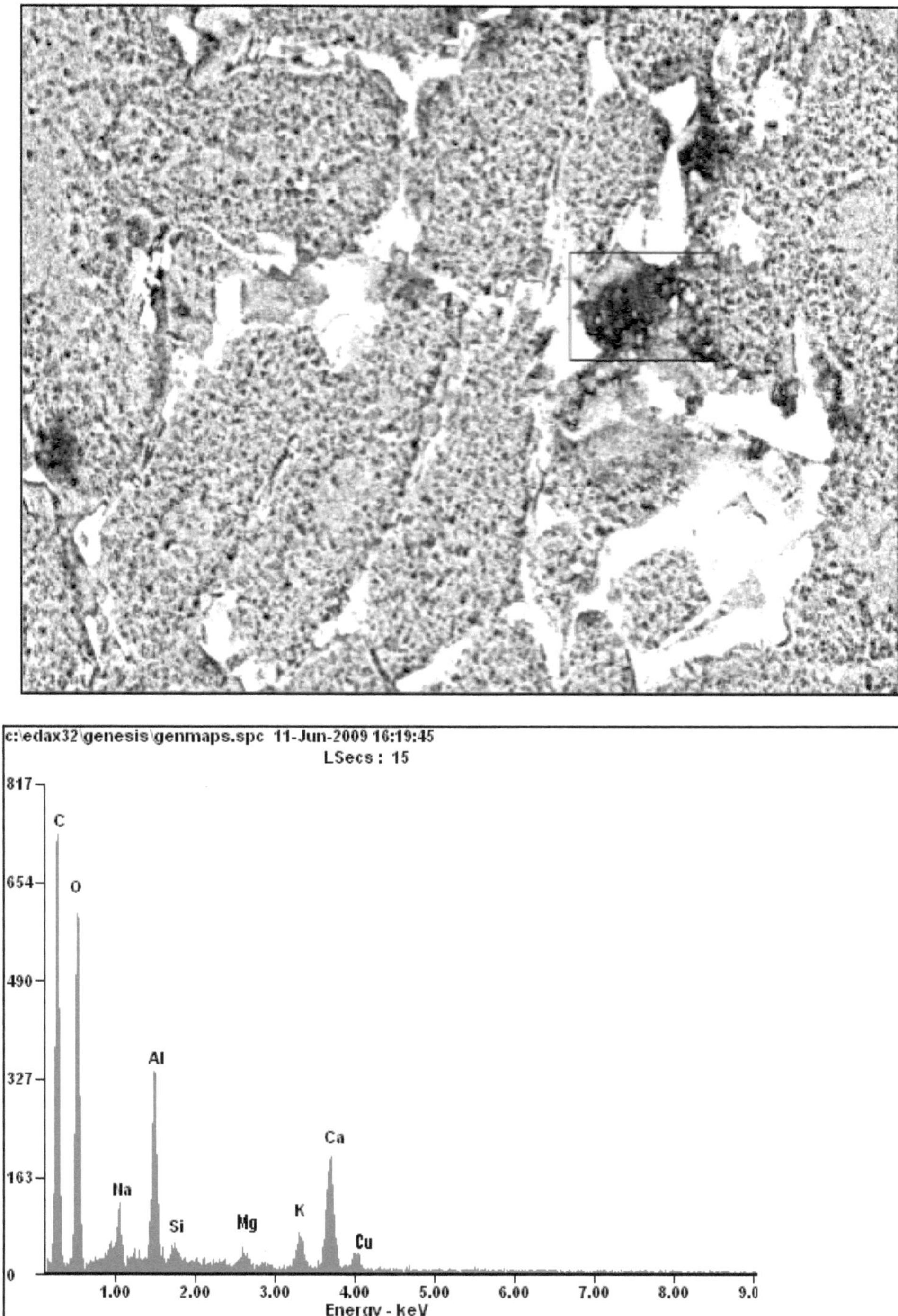

Figure 8.13: SEM/EDS of Microstructure of the Reinforced Al-Cu-Mg Alloy with 10wt per cent Bagasse Ash

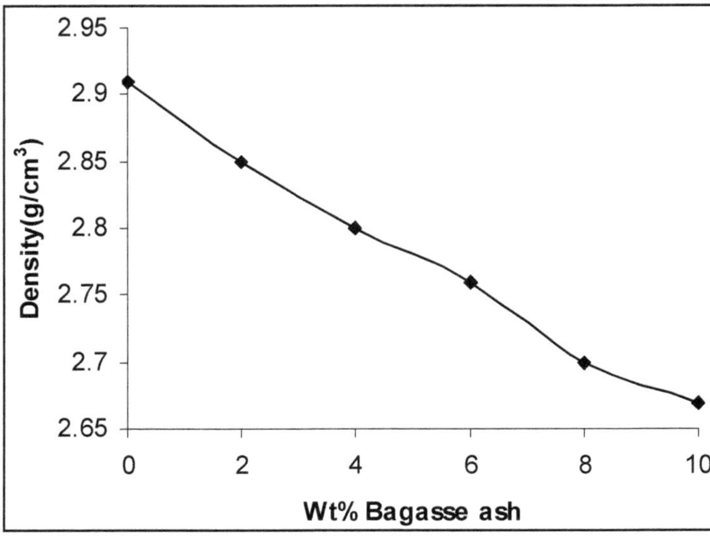

Figure 8.14: Variation of Density with Wt per cent Bagasse Ash

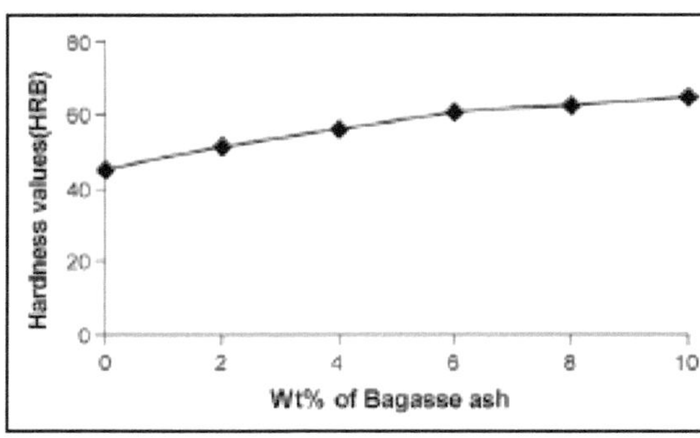

Figure 8.15: Variation of Hardness Values with Wt per cent Bagasse Ash

boundary of the ceramic particles by the difference in the thermo- expansion coefficient between the matrix and ceramic particles.

The yield and ultimate tensile strength increased with an increasing percentage of bagasse ash additions up to maximum values of 150.60 and 206.38N/mm2 at 8wt per cent of bagasse ash addition and then decreased to 145.50 and 168.50N/mm2 at 10wt per cent of bagasse ash addition (Figure 8.16).

The maximum values obtained for both ultimate tensile and yield strength in this composite at 8wt per cent bagasse ash are attributed to the uniform distribution of the bagaase ash particles in the microstructure (Figure 8.9) as compared with that of 10wt per cent of bagasse ash addition (Figure 8.10). It is clear that the inter-atomic spacing is smaller and the contribution of inter-particle distance to strain hardening arises from the fact that the space permitted a dislocation to maneuver round obstacles limit

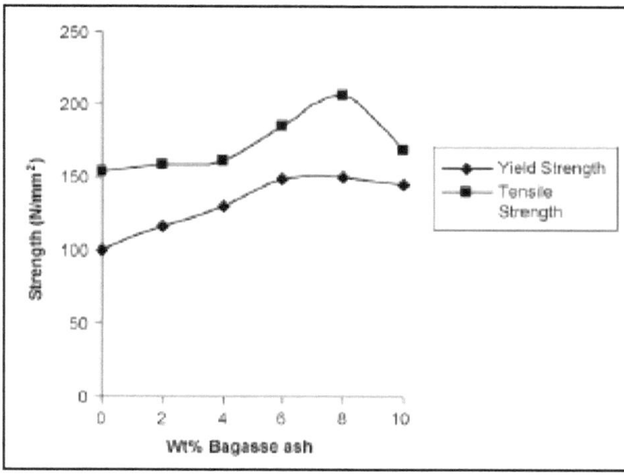

Figure 8.16: Variation of Strength with Wt. per cent Bagasse Ash

the yield stress as given by the formula: $\tau = \dfrac{Gb}{\lambda}$

Where τ = Yield stress, λ= Inter-particle distance, G= Shear energy of dislocation and b= Burger vector. This, therefore supports the fact that with smaller λ there is greater strengths than larger λ (Figures 8.9–8.10). These results show that Al-Cu-Mg/ 8wt per cent bagasae ash particles sample has 49.76 per cent and 34.25 per cent increase in yield and ultimate tensile strength over that of unreinforced Al-Cu-Mg alloy (A2009).

Compressive strength of the alloy and its composites are presented in Figure 8.17. It is clear that the compressive strength increased as the percentage bagasse ash particles increases in the alloy. This may be due to the hardening of the matrix by bagasse ash particles

The per cent elongation decreased with increase in weight fraction of bagasse ash particles (Figure 8.18).

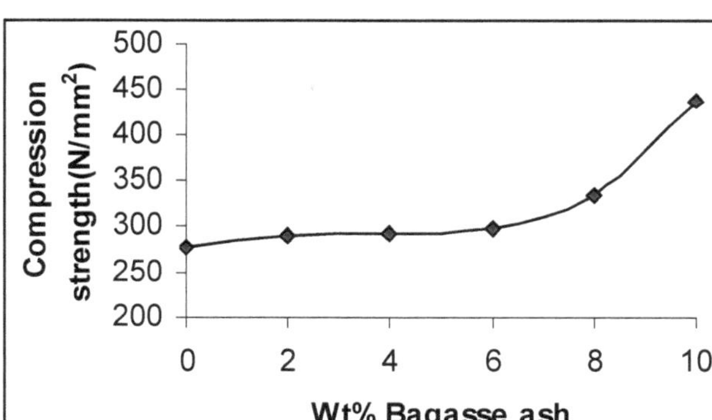

Figure 8.17: Variation of Compression Strength with Wt. per cent Bagasse Ash

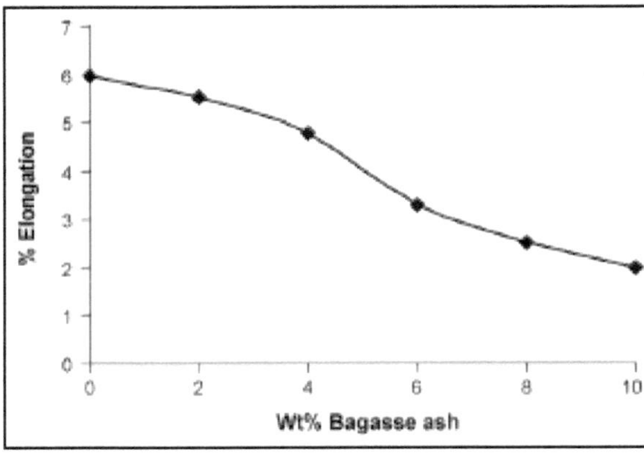

Figure 8.18: Variation of per cent Elongation with Wt. per cent Bagasse Ash

The steep reduction in the percent elongation of the composites could be attributed to the presence of particle clusters and weight fraction of second phase particles. This factor contributes to decrease in weight fraction of the ductile phase (weight fraction of the matrix phase) and thereby leads to decrease in percent elongation and also it is obvious that plastic deformation of the mixed metal matrix and the non-deformable reinforcement is more difficult than the metal matrix. The impact energy of the samples decreased as the percent of bagasse ash particles additions increase in the alloy (Figure 8.19).

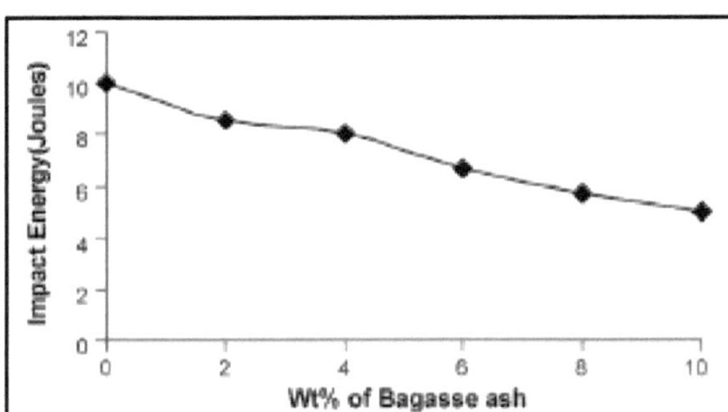

Figure 8.19: Variation of Impact Energy with Wt. per cent Bagasse Ash Additions

The brittle nature of the reinforcing materials (bagasse ash) plays a significant role in degrading the impact energy of the composite, since the unreinforced alloy and the alloy reinforced with 2wt per cent bagasse ash particles have the highest impact energy, indicating that they are the toughest of them all.

Figures 8.20 and 8.21 show the wear rate of matrix alloy and that of the composites.

From the Figures 8.20 and 8.21, the wear rate of the Al-Cu-Mg metal matrix reinforced with bagasse ash particles reduces with increasing reinforcement content and increase with increasing sliding velocity and applied load for dry sliding wear

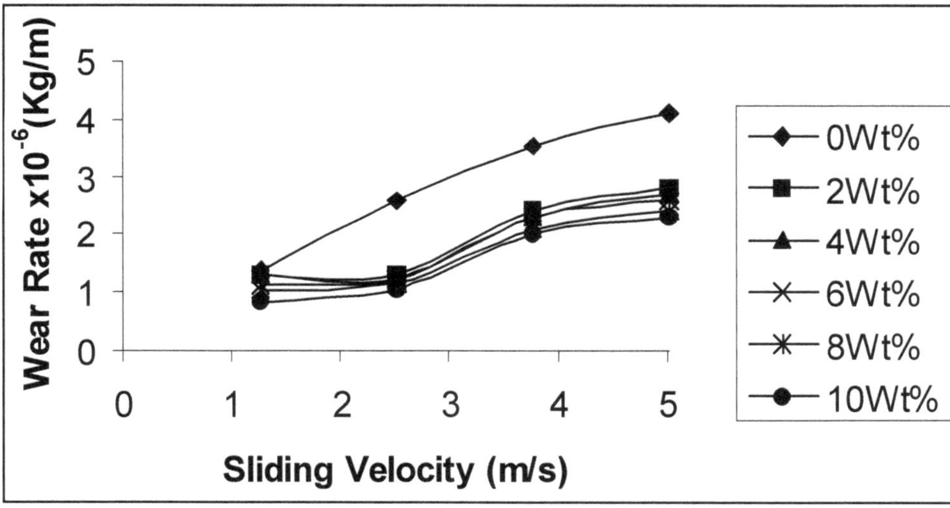

Figure 8.20: Variation of Wear Rate with Sliding Velocity at Constant Load of 20N

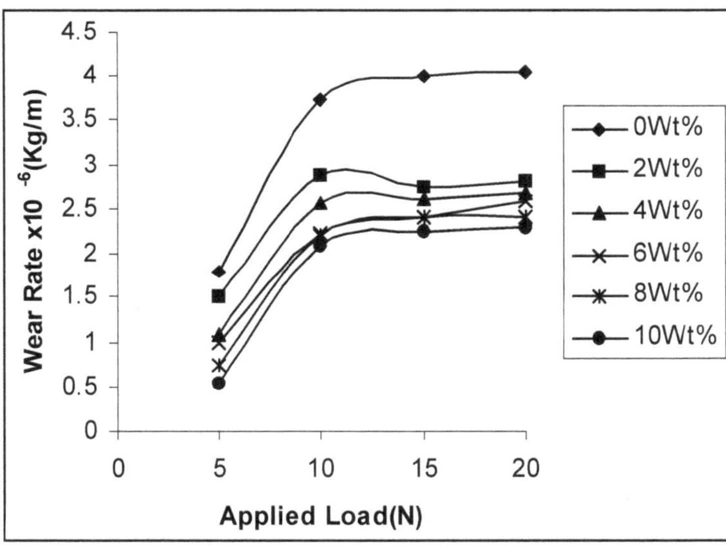

Figure 8.21: Variation of Wear Rate with Applied Load at Constant Speed of 5.02m/s

tests. However, the beneficial effect of the reinforcement on the wear resistance of the composites is observed to be the best at low loads, which reduces with increase in load applied. This observation is on par with one observed by Jacobson *et al.* (1985) who worked on silicon carbide reinforced aluminum alloy composites reported that the beneficial effect of particle reinforcement was gradually reduced with increasing load. Consequently, the effect of the reinforcement on th e composite wear resi stan ce is better for low loads. With higher loads contact temperatures become high and plastic deformation occurs with consequence of very high wear (Aigbodion and Hassan, 2007).

The decrease in wear rate of the composites may also be attributed to higher load bearing capacity of hard reinforcing material and better interfacial bond between the particle and the matrix reducing the possibility of particle pull out which may result in higher wear (Figures 8.8–8.13). This has been confirmed by Sato and Mehrabian (1976) and Hosking *et al.* (1992) for the aluminum matrix–Al2O3 system

Conclusions

Considerable success was recorded in the synthesis of Al-Cu-Mg alloy with bagasse ash additions. From the results obtained in this research, it can be concluded that:

1. The presence of the bagasse ash particles in the matrix alloy results in a much smaller grain size in the cast composites compared to the matrix alloy.
2. The addition of bagasse ash particles to Al-Cu-Mg alloy increases both the yield strength, ultimate tensile strength and hardness value up to a maximum values of 150.60, 206.38Nmm2 and 62.10HRB respectively at 8wt per cent bagasse addition. However, it is accompanied by a general reduction in impact energy and percentage elongation.
3. Bagasse ash particles can be used as a reinforcement material to improve the wear properties of the aluminium alloy.
4. For optimum service performances of this alloy, bagasse ash addition should be between 6-8wt per cent and not exceed 8wt per cent in order to develop better necessary properties. These composites can be used in area of wear resistance i.e in the production of pistons, cylinder liners, brake discs and bushing e.t.c. since aluminum composites used in production of these components has similar properties to that of Al-Cu-Mg/BAp composites.

Acknowledgements

The authors acknowledge with thanks the University of the Witwatersrand, Johannesburg, for allowing us to use their equipment. The laboratory assistance of Mr. Saka Ambali Abdulkareem is highly acknowledged as well. We also acknowledged the Department of Metallurgical and Materials Engineering, Ahmadu Bello University, Zaria for making their laboratory available for the work.

References

Aigbodion, V.S. and Hassan, S.B., 2007. Effects of silicon carbide reinforcement on Microstructure and properties of cast Al-Si-Fe/SiC particulate composites. *Journal of Materials Science and Engineering,* A447: 355–360.

Aigbodion, V.S., 2007. Particulate-strengthened of Al-Si alloy/alumino-silicate composite. *Materials Science and Engineering,* A460–461: 574–578.

Aigbodion, V.S., 2010. Potential of using Bagasse ash particle in Metal Matrix Composite. *PhD Dissertation,* Department of Metallurgical and Materials Engineering, Ahmadu Bello University, Samaru, Zaria, Nigeria(2010), pp 34–50.

Bienia, J., Walczak, M., Surowska, B. and Sobczaka, J., 2003. Microstructure and corrosion behaviour of aluminum fly ash composites. *Journal of Optoelectronics and Advanced Materials*, 5(2): 493–502.

Jacobson, Clyne, T.W., Liu, J. and Yao, C.K., 1985. Wear behaviour and microstructural changes of SiC-Al composite under unlubricated sliding friction. *Wear,* 184: 187–192.

Hosking, F., Folgar, F. and Wunderlin, R., 1982. Wear behaviour of SiC-Al composite. *J. Mater. Sci.,* 17: 477.

Naresh, P., 2006. Development and characterization of metal matrix composite using red Mud an industrial waste for wear resistant applications. *Ph.D. Thesis,* Department of Mechanical Engineering, National Institute of Technology Rourkela –769 008 (India) January, pp. 23–34.

Rajan, T.P.D., Pillai, R.M., Pai, B.C., Satyanarayana, K.G. and Rohatgi, P.K., 2007. Fabrication and characterization of Al-7Si-0.35Mg/fly ash metal matrix composites processed by dierent stir casting routes. *Composites Science and Technology*, 67, 3369–3377.

Sato, A. and Methrabian, R., 1976. Abrasive wear of aluminium matrix-Al_2O_3 System, *J. Metall. Trans.,* 7B: 443.

Chapter 9

Investigation of Austenitising Condition on the Tempering Characteristics of High-Strength Ferritic Ductile Iron

K.A. Bello, M. Abdulwahab, J.O. Gaminana and O. Salami*

Department of Metallurgical and Materials Engineering, Ahmadu Bello University Zaria, Kaduna State, Nigeria

ABSTRACT

There is a need to develop and improve the performance of engineering parts for various industrial applications. Recently, ductile iron has emerged as a major engineering material as it possesses attractive properties such as high strength with good ductility and toughness. One way to enhance these properties is by heat treatment technology which yields outstanding microstructure and mechanical properties that can be varied over a wide range by changing the heat treatment parameters. The present investigation is based on experimental studies to elucidate the influence of austenitising and tempering temperatures on the structure and properties of unalloyed ferritic ductile iron. In this study, some unalloyed ductile iron samples were partially austenitized in the in the two phase (a+l) region at intercritical temperature of 780°C for 1hr, quenched and then tempered over a range of temperature from 250°C to 650°C to obtain favourable structure and properties. For the purpose of comparison, some samples were also conventionally quenched from 900°C and tempered. Test samples were examined for tensile strength, ductility and toughness. Experimental results showed that intercritically treated and tempered samples have exhibited much toughness and ductility than in conventional tempered samples when tested at the

* *Corresponding Author:* E-mail: kabello@abu.edu.ng, bellkam@yahoo.com

same strength level. This is attributed to the lower carbon content in martensite formed during quenching from intercritical temperature. Thus, it is possible to reduce structural component weight, primarily in automotive industry, as the newly developed ductile iron is characterized with high strength that could permits weight reduction without sacrificing formability.

Keywords: *Tempered ductile iron, Tempering temperature, Austenitising temperature, Intercritical, Ferritic ductile iron.*

Introduction

The term cast iron refers not to a single material, but to a family of materials whose major constituent is mainly iron with important amount of carbon, silicon and other elements such as sulphur, phosphorus and manganese which are basically introduced intentionally or as consequence of the production process it is subjected to (Rajan, 1988). The presence of trace elements, the addition of alloying elements, the modification of solidification behaviour and heat treatment after solidification are used to change the microstructure of cast irons to produce the desired mechanical properties (Henkel *et al.,* 2002).

For a long time, gray cast iron was the only cast iron material available (Kraus, 1980). Ductile cast iron referred to as nodular or spheroidal graphite cast iron, is a recent member of the family of cast irons. Chemically, ductile iron is the same as gray iron and consists of iron-carbon-silicon alloy. The most important and distinguishing microstructural feature of all ductile iron is the presence of graphite nodules in eutectic, which acts as "crack arresters" and give ductility and toughness to ductile iron, which are superior to all other cast irons and equal to many cast and forged steels (Labrecque *et al.,* 1998 and Stasys *et al.,* 2008). Ductile cast iron provide a wide spectrum of mechanical properties that can be obtained either by altering certain processing variables or through various post heat treatments which present different and better combination of properties for application with special requirement (Raghavan, 1989 and Labrecque *et al.,* 1998). Because of these modification processes, ductile iron is now finding more applications as structural components in various industries such as automotive, agricultural and defence.

In the present day metallurgical practice, heat treatment has come to represent a whole class of techniques involving thermal, mechanical and chemical treatment by which the desire properties of structural materials are obtained to suit a variety of applications (Rajan *et al.,* 1988 and Trudel *et al.,* 1997). Tempering is one of the most important heat treatment processes that are applied to quenched steel and cast iron. The objectives of this process include reducing the brittleness of materials, improvement of toughness and ductility and also, reducing the problem of cracking (Krauss, 1980 and Vernik *et al.,* 1981).

Although ductile cast iron and steels are superficially similar but the high carbon and silicon level in ductile iron results in an important difference in their response to heat treatment (Trudel *et al.,* 1997 and Callister 1997). The high level of carbon in ductile cast iron can cause quench crack due to the formation of high martensite and/or the retention of metastable austenite (Hakan *et al.*, 2008 and Daber *et al.*, 2008). These undesired phenomenons, make the control of composition and austenitising temperature and quenching condition more critical in ductile iron. The transformation characteristics of ductile cast iron from austenitization temperature range have been investigated following tempering (Mehmet *et al.*, 2007). High austenitising temperatures increase the carbon content of the austenite but the bulk hardness is reduced due to the retained austenite and lower resultant martensite content (Daber *et al.*, 2008). Recently, it has been shown that the tensile strength and ductility can satisfactorily be optimized by critical combinations of austenitizing and tempering conditions (Hafiz *et al.*, 1997 and Kokatepe *et al.*, 2006). This study was therefore, undertaken to investigate the effect of partial and full austenitization on the tempering behaviour of ferritic ductile iron.

Materials and Methods

Melting of raw material employed to produce unalloyed nodular cast iron considered for this investigation was performed in an oil-fired rotary furnace (20kg capacity). The charge material consisted of pig iron, return steel scraps, graphitizing agent (ferro-silicon alloy) and inoculating agent (ferro-magnesium alloy). The melt was poured from 1420oC into standard 12.5mm Y- block sand moulds prepared in accordance with ASTM A-395 to ensure sound casting without defects like shrinkage for mechanical test samples. Tensile and izod test samples were machined and prepared from the bottom section of the Y blocks in order to avoid segregation effects and porosity present at the top part. The chemical composition of ductile cast iron produced for the present investigation is provided in Table 9.1.

Table 9.1: Chemical Composition (wt. per cent) of Unalloyed Auctile Cast Iron Used

C	*Si*	*Mn*	*S*	*P*	*Mg*	*Cr*	*Ni*	*Fe*
3.60	2.69	0.29	0.008	0.013	0.05	0.035	0.016	Bal.

Samples for investigation were initially heated at 950°C for 2 h and then cooled down in the furnace. This is to obtain a complete ferrite structure as the initial microstructure in all cases. Two batches of heat treatment operations were considered for this investigation. In the first batch, some samples were partially austenitized at 780oC for 1hr and then quenched in oil. The quenched samples were subsequently tempered at 250°C, 350°C, 450°C, 550°C and 650°C for 1h followed by cooling in air. The second batch of samples were processed by quenching from conventional austenitising temperature of 900°C and then tempered.

Tensile tests of tempered samples were performed on universal hydraulic testing machine with 50KN loading capacity and a cross-head speed of 1mm/min. The yield stress (0.2 per cent proof stress), the ultimate tensile stress and elongation at failure were measured. The tempered impact specimens were tested using izod impact testing machine. All the tests were carried out at room temperature and minimum of two samples were tested for each heat- treated condition.

Results and Discussion

The mechanical properties, *i.e.* yield strength, tensile strength, hardness, percentage elongation and impact toughness are measured as a function of austenitizing and tempering temperatures. In this investigation, specimens were austenitized at intercritical temperature of 780°C and then tempered in the tempering temperature of 250-650°C at interval of 100°C. Some of the specimens quenched from 900°C were also tempered for comparison. As expected, the ferritic ductile irons responded favourably to the heat treatment conditions and the results obtained are presented as shown in Figures 9.1–9.5.

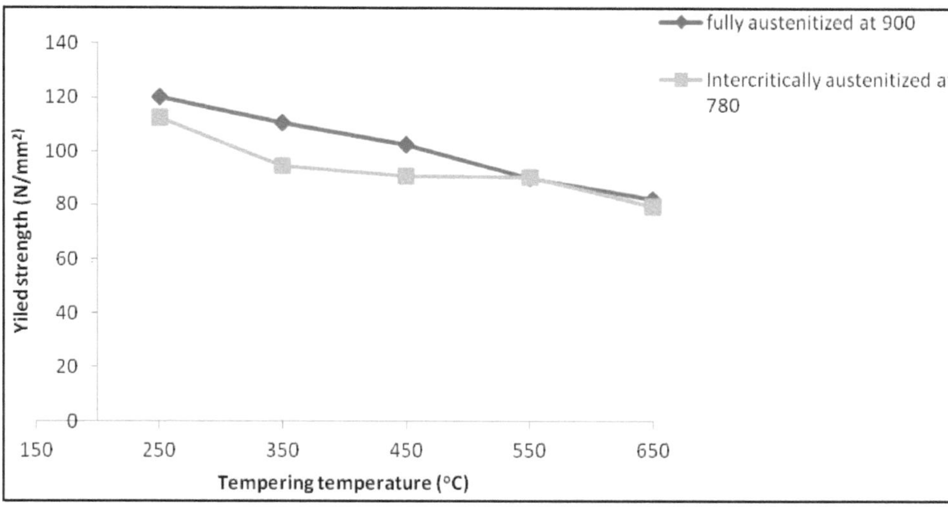

Figure 9.1: Variation of Yield Strength as a Function of Tempering Temperature for Austenitizing Condition of 780 °C and 900 °C

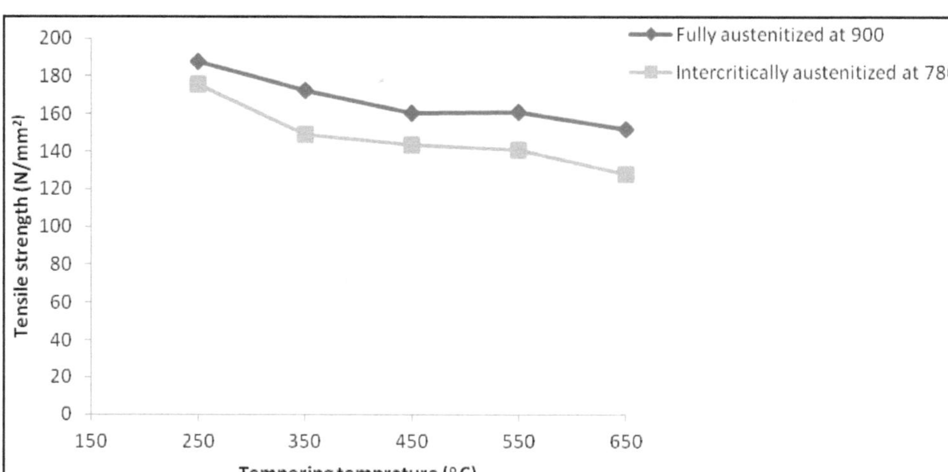

Figure 9.2: Variation of Tensile Strength as a Function of Tempering Temperature for Austenitizing Condition of 780 °C and 900 °C

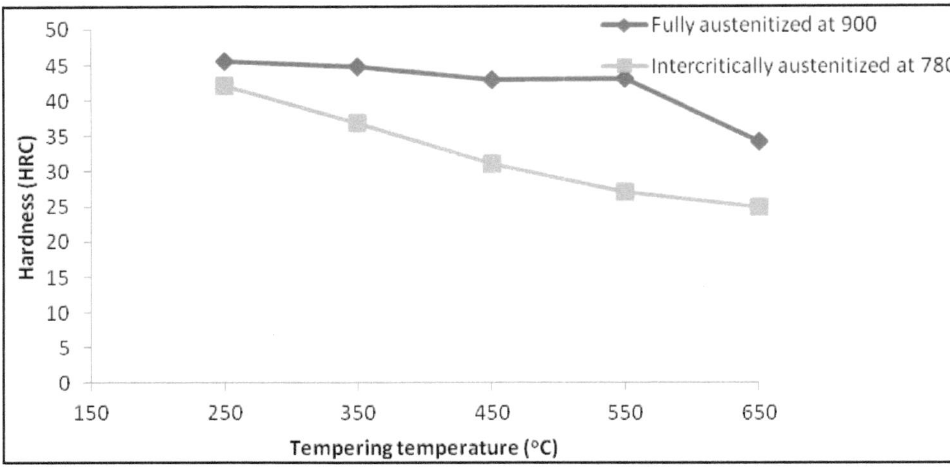

Figure 9.3: Variation of Hardness as a Function of Tempering Temperature for Austenitizing Condition of 780 °C and 900 °C

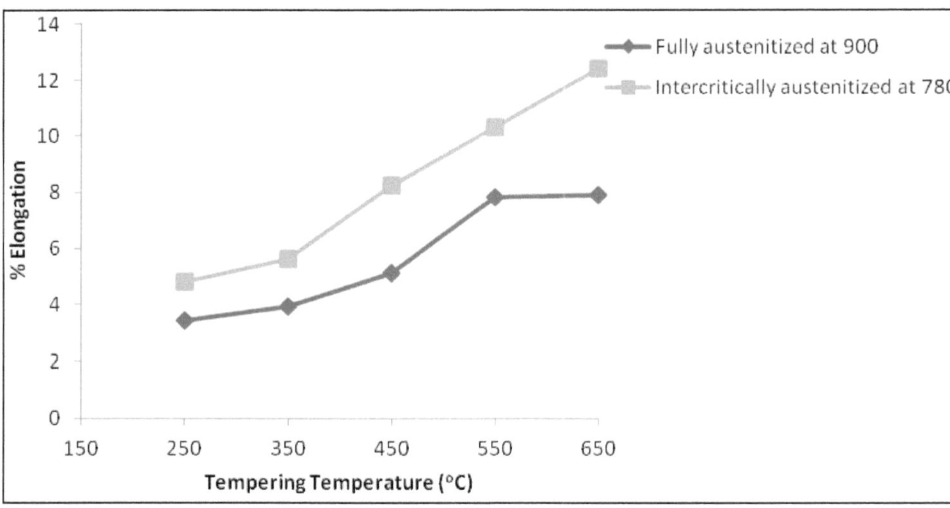

Figure 9.4: Variation of Elongation as a Function of Tempering Temperature for Austenitizing Condition of 780 °C and 900 °C

The plots of tensile strength, yield strength and hardness versus tempering temperatures demonstrates that the properties decrease linearly with increasing tempering temperature for the samples austenitized at higher austenitizing temperature of 900°C (Figures 9.1–9.3). However, the lower (780°C) austenitizing temperature does not exhibit the same behaviour as slight reduction in these properties was obtained for the partially austenitized samples as tempering temperatures decreases. This leads to the fact that tempering at lower tempering temperatures will result in lower hardness and strength as the tempered structure containing less of retained austenite and pro-eutectoid ferrite that was formed from incomplete austenitization of the samples.

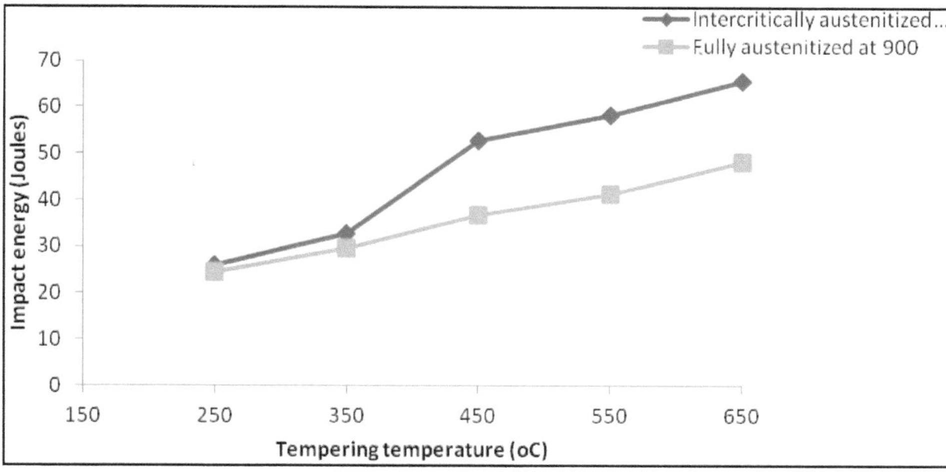

Figure 9.5: Variation of Impact Toughness as a Function of Tempering Temperature for Austenitizing Condition of 780 °C and 900 °C

Figures 9.4 and 9.5 report influence of tempering temperature on variation of ductility and toughness of ferritic ductile iron processed by intercritical and conventional austenitizing temperatures. The figures clearly indicate that both the percentage elongation and impact toughness significantly improve with increased tempering temperatures from 250-650 °C in all the tested samples. However, it is noticed that ferritic ductile iron samples tempered from lower austenitizing temperature of 780 °C exhibit higher values compared to other tempered from quenching temperature of 900 °C. This behaviour can be explained to be attributed to the large volume fraction of pro-eutectoid ferrite resulting from the incomplete austenitization of the tested samples. Also, this might be due to the fact that during partial austenitization, a dual phase of ferrite and martensite was produced on quenching which later transform to ferrite and tempered martensite microstructure. This unique combination of ferrite and tempered martensite is probably responsible for the higher ductility obtained for the partially austenitized and tempered ductile iron samples as compared to the conventional tempered condition.

Conclusions

1. Ductile iron samples that were partially austenitized, quenched and tempered exhibited the best combination of ductility and fracture toughness than the conventionally quenched and tempered conditions, at the same strength level. This increased in value is attributed to the role of proeutectoid ferrite and tempered martensite.
2. Within the tempering temperature range of 250-650°C, appreciable decrease in yield and tensile strength of the conventional tempered samples, whereas it remains almost unchanged for the intercritical austenitized, quenched plus tempered samples.

3. Lower austenitizing temperature leads to a considerable reduction in hardness and strength in the tempered samples but characterized with much higher toughness and ductility when compared with the conventional treated samples. Hence, samples quenched from intercritical temperature and tempered are becoming an advanced candidate material for automobile applications where good combination of strength, toughness and formability is a requirement.

References

Callister, W.D., 1997. *Materials Science and Engineering: An Introduction.* John Wiley and Sons Inc., New York, pp. 256–324.

Daber, K.S. and Ravishanka, P., 2008. Influence of austenitising temperature on the formation of strain induced martensite in austempered ductile iron. *Journal of Material Science*, 43: 4929–4937.

Hafiz, M., 2001. Tensile properties and fracture of ferritic SG-iron having different shell structure. *Z. Metallkd*, 11: 1258–1261.

Hakan, G., Melika, O. and Mehmet, E., 2008. The evaluation of structure-property relationships in the dual matrix ductile iron by magnetic barkhausen noise analysis. *17th World Conference on Nondestructive Testing*, Shanghai, China.

Henkel and Pense, 2002. *Structure and Properties of Engineering Materials*, 5th Edn. McGraw-Hill Publishers, America, pp. 150–358.

Kocatepe, K., Cerah, K.M. and Erdogan, M., 2006. Effect of martensite volume fraction and its morphology on the tensile properties of ferritic ductile iron with dual matrix structure. *Journal of Material Process Technology*, 178: 207–223.

Krauss, G., 1980. *Principles of Heat Treatment of Steel*, ASM, p. 1–6.

Labrecque, C. and Gagne, M., 1998. Review ductile iron: Fifty years of continuous Development. Rio Tinto Iron and Titanium Inc., technology Tracy, Quebee, Canada, pp. 343–375.

Mehmet, E., Volkan, K. and Bilge, D., 2007. Transformation characteristics of ductile cast iron austempered from intercritical austenitising temperature range. *Journal of Material Science*, Spinger Netherland, 44: 1573–1598.

Raghavan, V., 1989. *Physical Metallurgy: Principle and Practice.* Prentice-Hall of India Ltd., New Delhi, pp. 97–98, 221–236.

Rajan, T.V., Sharma, C.P. and Ashok, S., 1988. *Heat Treatment: Principle and Techniques.* Prentice-Hall of India Ltd., New Delhi, pp. 96–116, 142–50.

Stasys, B., Arvydas, V. and Gintautas, Z., 2008. Relation between section thickness, microstructure and mechanical properties of ductile iron castings. *Material Science*, 14(2): 115–118.

Trudel, A. and Gagne, M., 1997. Effect of composition and heat treatment parameters on the characteristics of ductile irons. Rio Tinto Feret Titane Inc., Technology, Sorel Quebee, Canada, pp. 289–297.

Vernik, E.D., 1981. *Iron Casting Handbook*. Iron Casting Society, pp. 111.

Chapter 10

Challenges and Opportunities of Using Composite Materials in Automobile Industry

S.B. Hassan[1], V.S. Aigbodion[1] and C.U. Atuanya[2]*

[1]Department of Metallurgical and Materials Engineering, Ahmadu Bello University, Samaru, Zaria, Nigeria
[2]Metallurgical and Materials Engineering Department, Nnamdi Azikiwe University, Awka

ABSTRACT

Composite materials have emerged as a necessity rather than compulsion in the present era of industrial applications. This scenario has caused the industry to look for better alternate materials which come at affordable prices to maintain the requisite price margin, besides considering the depleting natural resources of conventional materials. However, apart from offering better advantages in terms of lightweight, wear resistance, ease of maintenance and better environmental protection, also poses few disadvantages in terms of processing cost and the choice of material. This paradigm shift can translate only after extensive evaluation of available materials at the design stage aided with required experimental tests. Experimental procedures needs to be extensive as the material properties could be unpredictable in some circumstances. This paper discusses the opportunities available to discover areas of avenues in automobile industry in particular with brake system, engine parts to replace the primitive metallic components with metal matrix composite materials. A practical application of metal matrix composite in the automobile industry is presented.

Keywords: *Automobile industry, Composite and Metal matrix.*

* *Corresponding Author:* E-mail: aigbodionv@yahoo.com

Introduction

The extravagant achievements of basic science, thus far unheard-of integration of science and technology have become the catalysts of the changes occurring in our life, and this applies largely to structural and functional materials which also make up the material world surrounding us. This paper discusses about the opportunities and challenges faced in replacing conventional materials by composite materials. This introduction of an alternative to conventional metal materials thus needs to be "environment friendly".

Metal matrix composites (MMCs) have been used commercially in the automotive market for nearly 20 years. Properties of interest to the automotive engineer include increased specific stiffness, wear resistance, and improved high-cycle fatigue resistance (Aigbodion, 2007a). While weight savings is also important in automotive applications, the need for achieving performance improvements with much lower-cost premiums than tolerated by aerospace applications drives attention toward low-cost materials and processes (Aigbodion, 2007b).

Trends in the Development of Composite Materials

In the early 1900's, scientists increased their experimentation and development of innovative new materials. Invented originally by accident by Owens- Corning ("Fiberglass") research was then encouraged heavily by the defense industry during World War II. There is also a cultural change going on in the automobile market. The market is expressing new expectations which, is beyond price, and, involve a requirement for performance levels with due emphasis on the environment and sustainable development (Harrigan, 1990).

Over the last 15 years the utilization of metal composites has superseded that of aluminium or steel. Figure 10.1, shows the growth in world consumption of composites in comparison with conventional materials like aluminium and steel (Harrigan, 1990). Suggestive that the annual growth rate of metals composites being at forefront. Thus, offering many more creative design options calling for innovation. Though Composite manufacturing costs are higher than costs of standard materials such as steel or

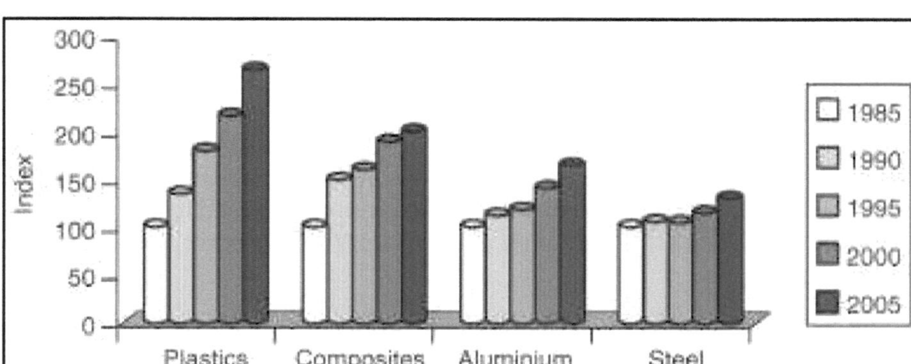

Figure 10.1: Evolution of World Consumption

aluminum. Actually, the composite solution is always a technological leap for designers. However, the benefit of composite materials over alternative solutions has to be evaluated at the time of design along with the required testing. Unlike composites, standard materials (steel or aluminum) seem to be reassuring solution because their technical performances are known and on file, and their wear performance is predictable. Also, these standard materials are regularly being improved (lightweight, special metal processing) (Hamajima *et al.*, 1990).

Metals Composite Automobile Applications

Composite materials are traditionally designed for use as structural materials. With the rapid growth of the automobile industry, Composite materials are finding more and more applications in automobile. There has been successful application in several automotive applications in which the combination of properties and cost satisfied a particular need. As a result of these successful applications, the 1999 total world market of MMCs for ground transportation (automotive and rail) amounted to 1566 metric tonnes, valued at over $7 million (Hassan and Aigbodion). Some of these applications are described subsequently, emphasizing the innovations that have led to MMCs insertion and the system-level benefits that have been obtained. Some of the areas of application in the automobile sector are as follow:

Connecting Rods

Other drive train components, and particularly the connecting rod, have been a focus of development (Donomoto, 1983). By reducing the mass of the connecting rod/ piston assembly, the objectionable secondary shaking forces that can develop, particularly in smaller engines, can be reduced. In addition, lower reciprocating loads should lead to lower loads on the crankshaft and lower friction losses, and increased fuel. Plate 10.1, shows the photomacrograph of connecting rod produced with MMCs.

Brake System

Aluminum-based MMCs offer a very useful combination of properties for brake system applications in replacement of cast iron. Specifically, the wear resistance and high thermal conductivity of aluminum MMCs enable substitution in disk brake rotors and brake drums, with an attendant weight savings on the order of 50 to 60 per cent. Because the weight reduction is unsprung, it also reduces inertial forces, providing an additional benefit in fuel economy. In addition, lightweight MMC rotors provide increased acceleration and reduced braking distance. It is reported that, based on brake dynamometer testing, MMC rotors reduce brake noise and wear, and have more uniform friction over the entire testing sequence compared to cast iron rotors (Aigbodion, 2010). Plate 10.2, shows the photomacrograph of brake rotors produced with MMCs.

Pushrods

The aluminum MMCs pushrods provide 25 per cent higher bending stiffness and twice the damping capacity relative to 4340 steel pushrods (Ejiofor and Reddy, 1997). The MMC density is less than half that of steel, so that engine speed can be

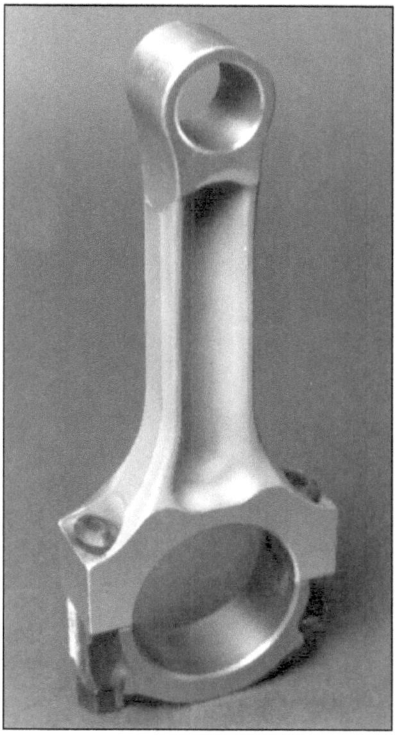

Plate 10.1: Prototype Aluminum MMC Connecting Rod

Plate 10.2: Aluminum MMCs Brake Rotors

increased by an average of 250 to 400 rpm before valve bounce begins. In addition, it has been projected that the use of aluminum MMCs pushrods could provide an extension in spring life of up to 600 per cent as a result of the improved damping characteristics of the MMCs (Ejiofor and Reddy, 1997). Because valve train spring sets are generally a significant expense, this benefit provides a value far in excess of the initial acquisition cost. Plate 10.3, show the photomacrograph of Pushrods produced with MMCs.

Challenges

Evaluating composite performance, engineering the equipment and procedures for product characterization, and validating the recycling technologies for composite materials are the three major challenges for research and development in the composite materials sector. Recycling is one of the key issues of the composite industry. A major challenge that faces the automobile industry over the coming years is how to deal with production and end–of–life waste (Bienia *et al.*, 2003).

Also high cost of current MMCs, interfacial reaction and high density of most common ceramic reinforcement compared to aluminum has inhibited production on a large industrial scale, for example in the automotive industry. In attempts to overcome this limitation, several research and development (R&D) programmes (Naresh, 2006)

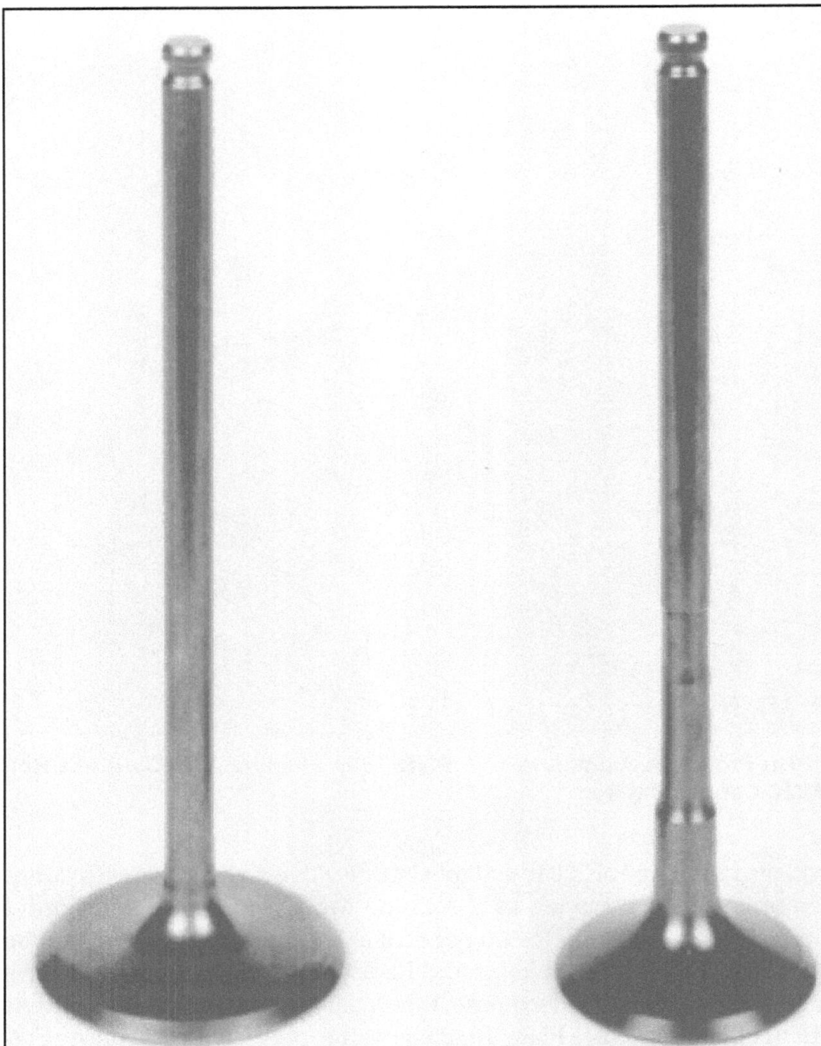

Plate 10.3: The Photomacrograph of Pushrods

were focused on the reinforcement of aluminum-based MMCs using low cost industrial waste by-products as the reinforcement particulate. This could reduce the cost and the weight of energy intensive metals for potential applications in engineering components for a new generation of vehicles. However, the future commercialization of these materials in automotive industry will be also influenced by the ability to recycle the scrap and the final components when its useful life is over.

Thus researchers have worked out separately to reinforce SiC, Al_2O_3 (*i.e.* carbides, nitrides and oxides) TiB_2, Boron and Graphite into the Aluminium matrix to achieve different properties and they are expensive and of high density (Hassan and Aigbodion, 2010) (Figure 10.2).

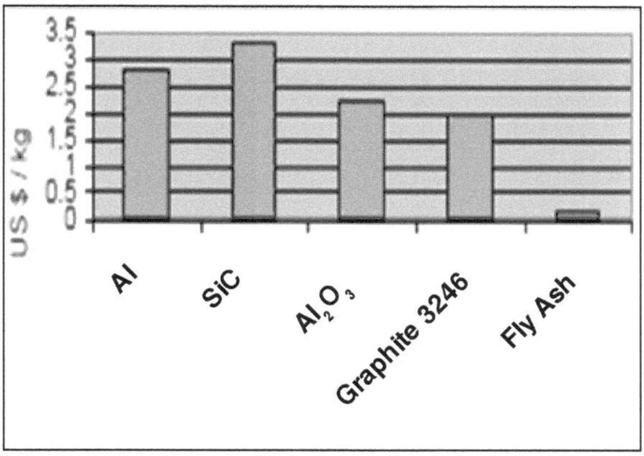

Figure 10.2: Prices of Materials Commonly Used in making MMCs

The ever-increasing demand for low cost reinforcement stimulated the interest towards production and utilization of using bye-products from industry as reinforcement since they are readily available or are naturally renewable at affordable cost. Aigbodion (2007a) have used alumino-silicate (Kankara clay) in reinforcing Al-Si alloy, Bienia *et al.* (2003) used fly ash in reinforcement of aluminium matrix, Naresh (2006), worked on the development and characterization of metal matrix composite using red mud an industrial waste for wear resistant applications, they all reported good dispersion and recovery of the particles in the composites castings.

Recently Aigbodion (2010) and Hassan and Aigbodion (2010) has found out the suitability of using bagasee ash(sugarcane waste) in the development of metal matrix for the automobile industry in the study. The authors found out that, the developed composites can be used in the development of lower weight auto-components and also in area were wear resistance is vital i.e in the production of pistons, cylinder liners, brake discs and bushing etc. some of the foundings from the authors are shown below:

The authors concluded that for further development of this developed materials to industrial status, the federal government of Nigeria and national automotive council of Nigeria are called upon.

Opportunities

Metal matrix composite offer the following opportunities in the automobile industry:

Design Requirements

1. Design freedom : realization of all shape and size parts
2. Integration of several functionalities by using the property versatilities such as structural and other additional properties : damping, shock and noise absorption and heat insulation
3. The possibility of selective reinforcement in the direction of the stresses by selecting particular composites or by part drawing.

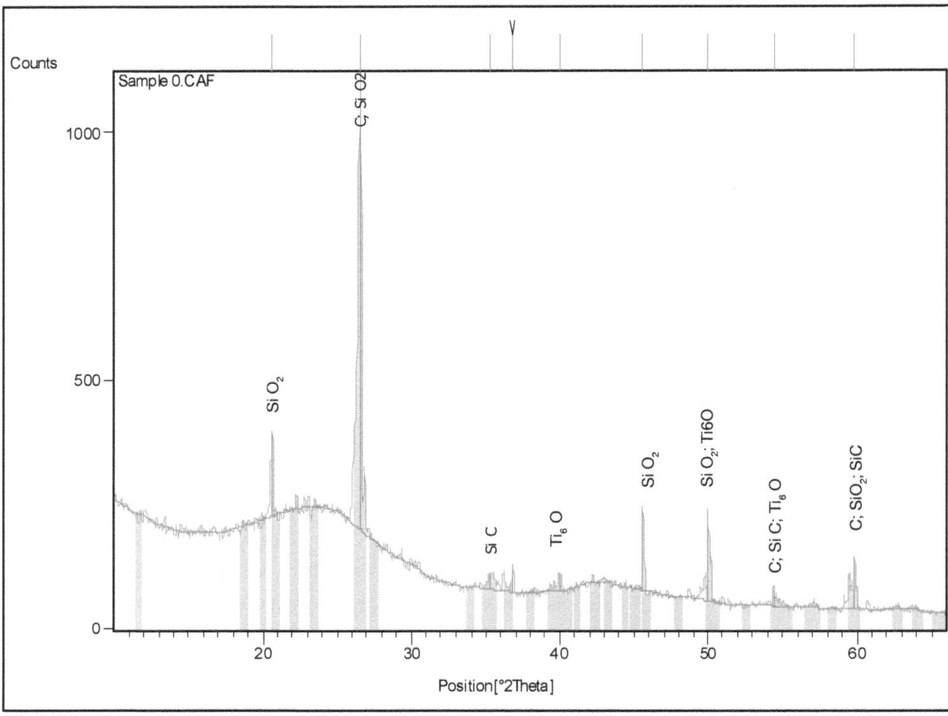

Figure 10.3: XRD Pattern of the Bagasse Ash

Figure 10.4: Photomicrograph of the Bagasse Ash

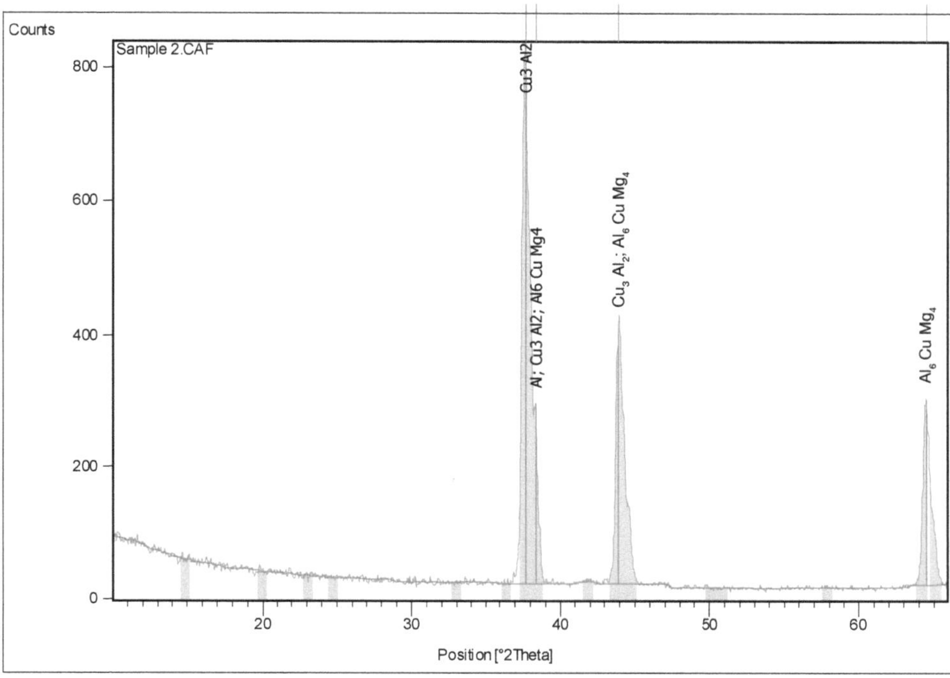

Figure 10.5: XRD Pattern of the Aluminium Alloy (A2009)

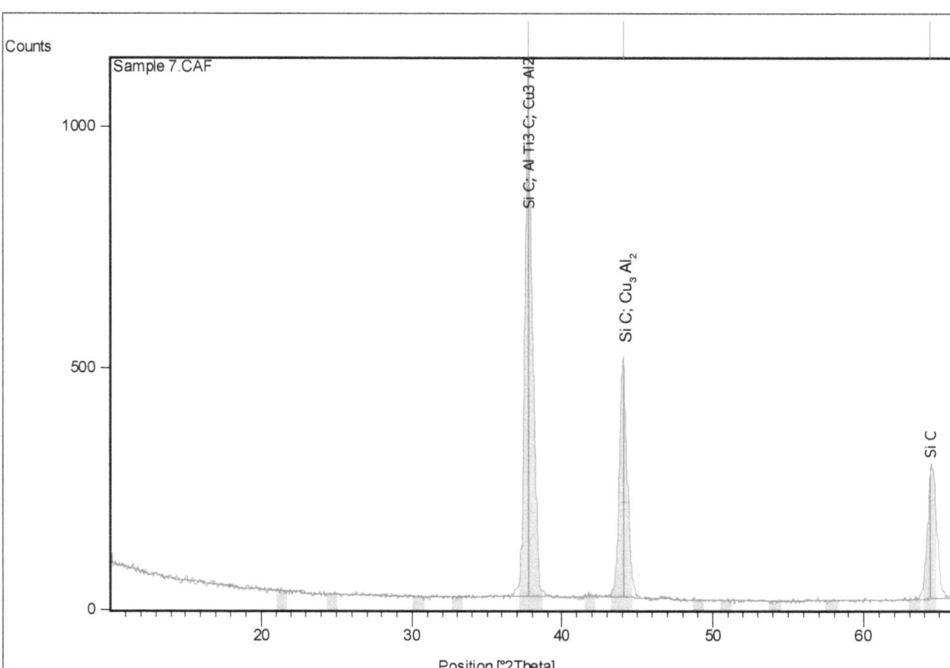

Figure 10.6: XRD Pattern of the Aluminium Alloy (A2009) Reinforced with 8wt per cent Bagasse Ash

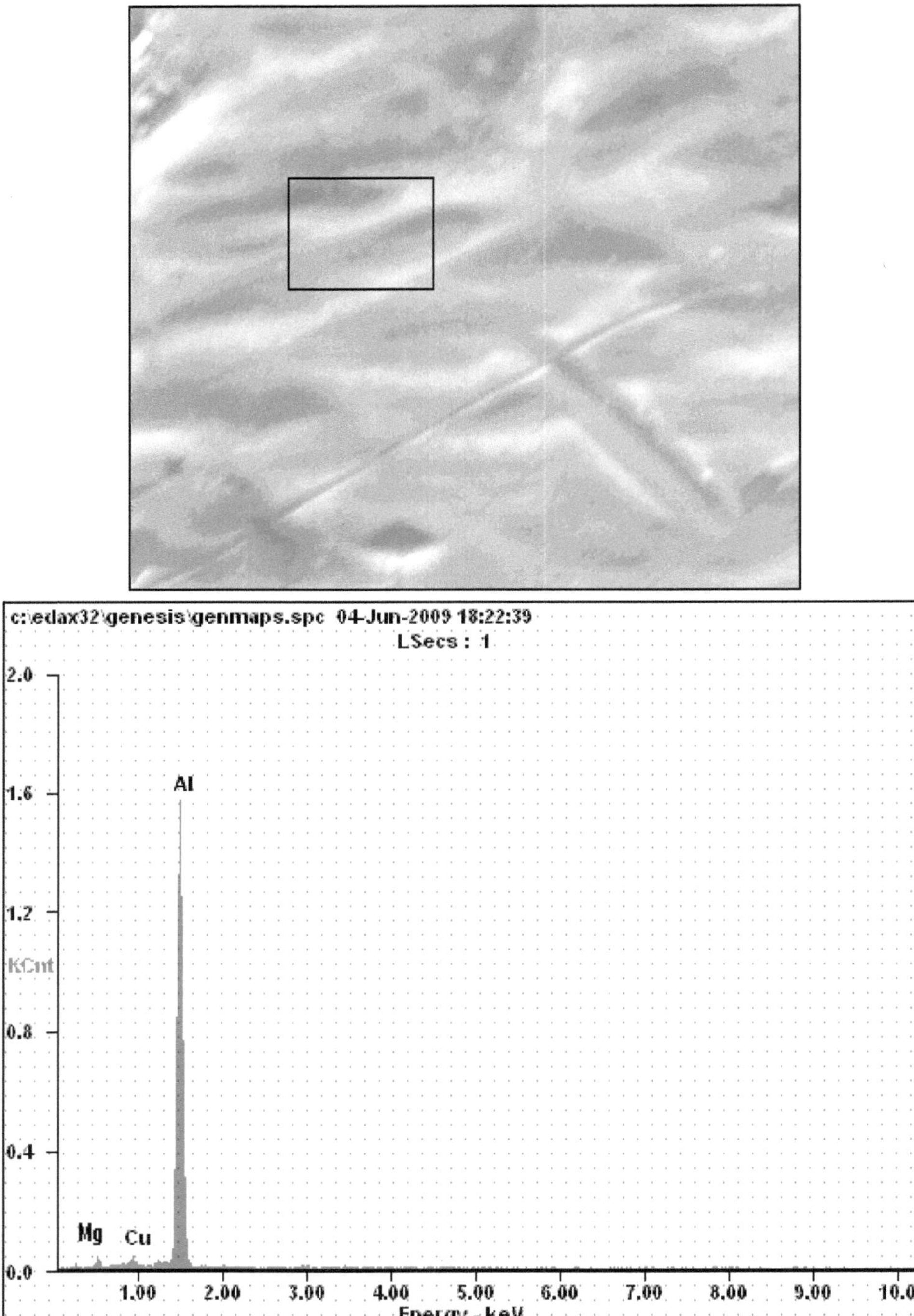

Cu_3Al_2 and Al_6CuMg_4 in α-Al

Figure 10.7: Photomicrograph of SEM Microstructure of the Unreinforced Alloy/EDS Spectrum (x400). The structure reveals the presence of Cu_3Al_2, Al_6CuMg_4 phases in α-Almatrix(white).

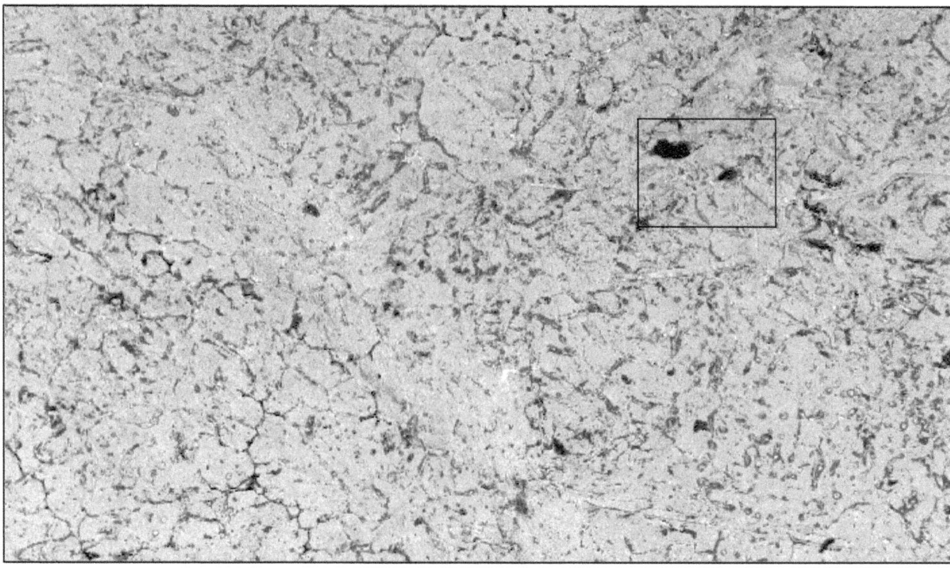

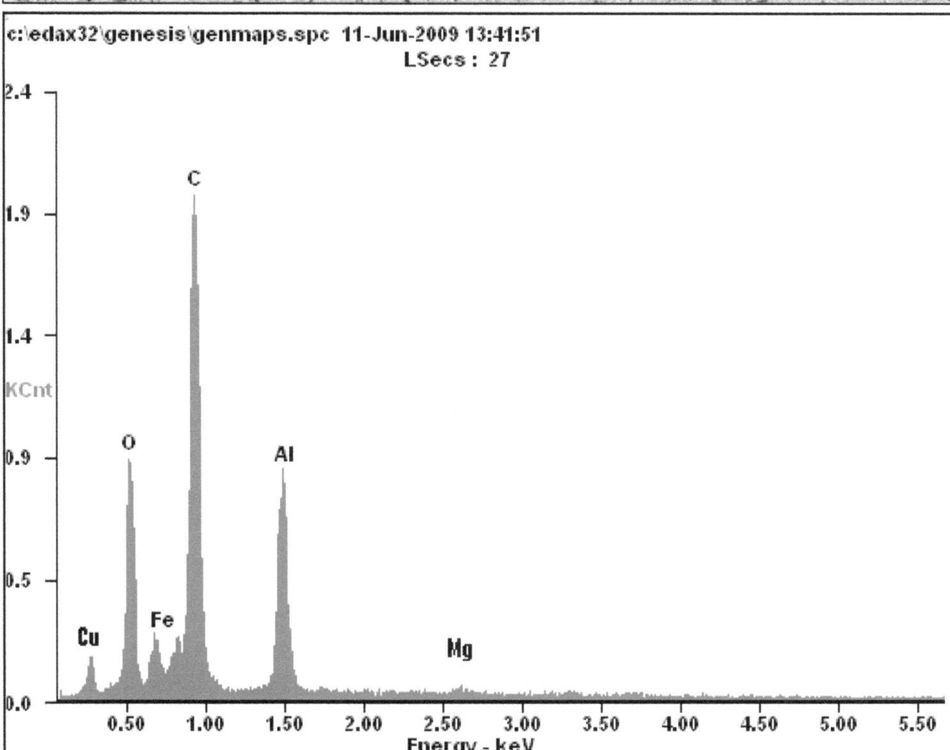

BAp

Figure 10.8: Photomicrograph of SEM Microstructure of the Reinforced Al-Cu-Mg Alloy with 4wt per cent Bagasse Ash/EDS Spectrum) (x400). The structure reveals the dissolution of the Cu_3Al_2, Al_6CuMg_4 phases and slight distribution of bagasse ash (Black) in α-Almatrix (White).

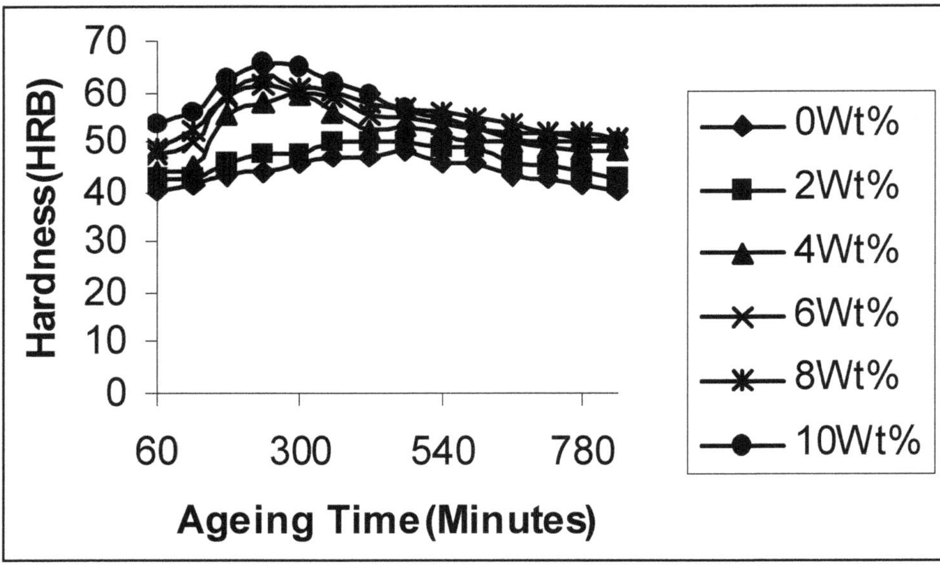

Figure 10.9: Variation of the Hardness Values with Ageing Time

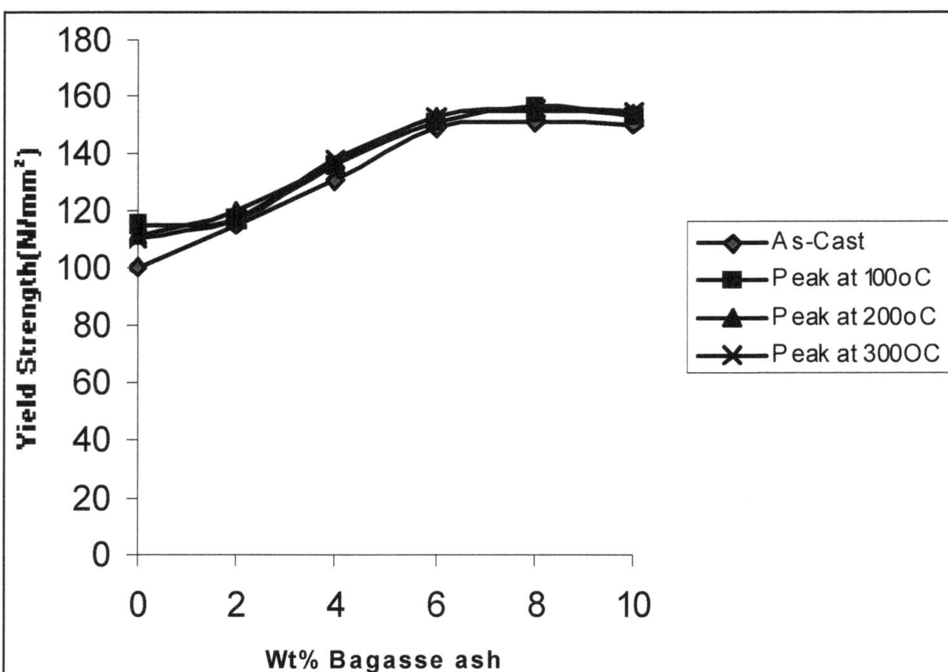

Figure 10.10: Variation of the Yield Strength with Bagasse Ash Addition

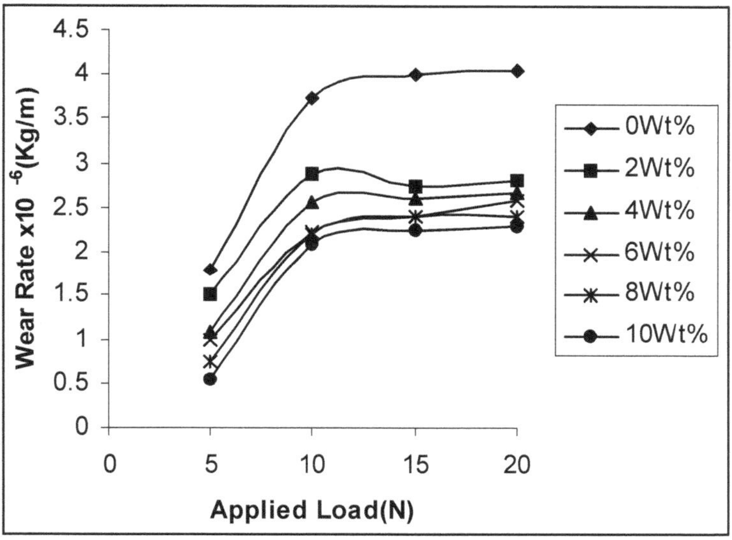

Figure 10.11: Variation of Wear Rate with Applied Load at Constant Speed of 5.02m/s

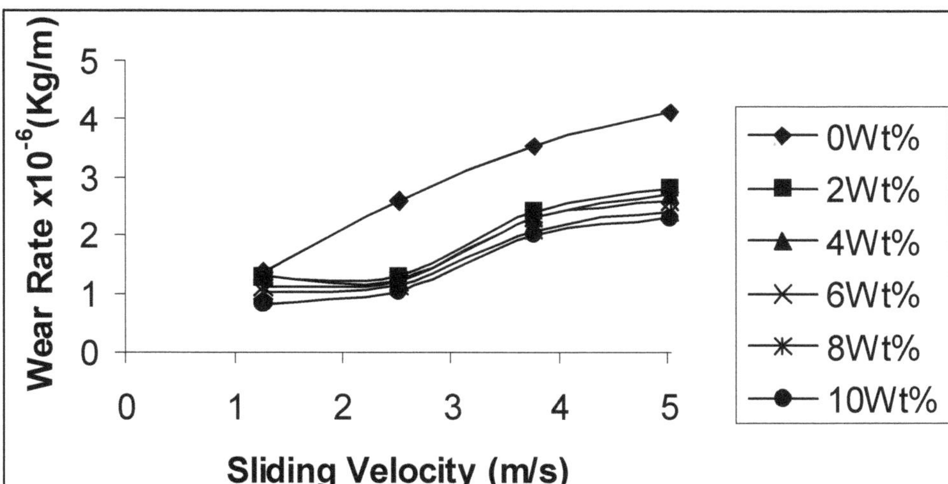

Figure 10.12: Variation of Wear Rate with Sliding Velocity at Constant Load of 20N

Technical Requirements

1. Solidity, reliability and permanence of the parts in the face of increasingly harsher environments and higher temperatures
2. Ease and reduction of maintenance
3. Slow damage propagation

Environmental Requirements

1. Weight reduction thanks to good mechanical properties combined with low density.
2. The reduction of suppression of the periodic painting of metals contributes to a reduction in pollution. There were several prospects for application in automobile industry examples:
   - ✰ Miniaturizing of design, reduction of used material, higher reliability
   - ✰ Long-term improvement of efficiency and elongation of life time period

A growing market has been established for MMCs in the automotive industry. Important applications have resulted when performance and weight benefits overcome the added component cost. Performance benefits have been obtained from reduced weight, improved wear resistance, low coefficient of thermal expansion, and good thermal properties. However, high volume process and material innovations are needed before MMCs are widely applied to the automobile across all product offerings. Continued pressure in the ground transportation industry for lower weight to improve environmental emissons and fuel efficiency is expected to generate growth in the existing application areas, as well as new opportunities for MMCs.

All of these ideas can not be put into reality within days or months; it will take a middle-term to long-term period from the idea to research work, over first prototypes and to industrial production and the application by the consumers.

Conclusios

The opportunities available to discover areas to replace the primitive metallic components with composite materials of avenues in automobile industry in particular is thus envisaged. With the present scenario calling for enormous amount of metal consumption in various industrial applications, this gradual trend is driving the natural resources to deplete. With an ambition to improve the environmental profile of automobile industry. metal matrix composite materials have been looked into as an alternate. Ability to solve the engineering challenges of today and tomorrow is still being driven by the cost and availability.

One needs to overcome this challenge to achieve the required degrees of success. Our understanding of how materials interact, how they can be reliably integrated with combinations in functional requirements have to be explored. The challenges, as well as the opportunities, will be in the areas of materials development, synthesis, fabrication, integration, characterisation and validating the recycling technologies, which of course aids capitulate lot of potential in Innovating new ideas. Thus, the automobile industry has to be more proactive in implementing environmental progress plans which will reduce the depletion of non renewable resources like steel.

References

Aigbodion, V.S., 2007a. Development of Al-Si-Fe/SiC particulate composite as advanced materials for engineering applications. *M.Sc Thesis,* Department of Metallurgical Engineering, Ahmadu Bello University, Zaria Nigeria, pp. 1–56.

Aigbodion, V.S., 2007b. Particulate-strengthened of Al-Si alloy/alumino-silicate composite. *Materials Science and Engineering*, A460–461: 574–578.

Aigbodion, V.S., 2010. Development and characterization of using bagasse ash particle in metal matrix composite. *Ph.D. Disseration,* Department of Metallurgical and Materials Engineering, Ahmadu Bello University, Samaru, Zaria, Nigeria, pp. 34–50.

Bienia, J., Walczak, M., Surowska, B. and Sobczaka, J., 2003. Microstructure and corrosion behaviour of aluminum fly ash composites. *Journal of Optoelectronics and Advanced Materials*, 5(2): 493–502.

Donomoto, A., 1983. SAE Paper 830252, Society of Automotive Engineers.

Ejiofor, J.U. and Reddy, R.G., 1997. Developments in the processing and properties of particulate Al-Si composites. *JOM Publication of Minerals, Metals and Materials Society*, 49(11): 1–37.

Harrigan, W.C., 1994. *Handbook of Metallic Composites,* S. Ochiai (1990) Ed., Marcel Dekker, Inc., New York, p. 759–773.

Hamajima, K.A. Tanaka, and Suganama, T., 1990. *JSAE,* 11: 80–84.

Hassan, S.B. and Aigbodion, V.S., 2007. Experimental correlation between varying silicon carbide and hardness values in heat-treated Al-Si-Fe/SiC particulate composites. *Materials Science and Engineering* A,454–45: 342–348.

Hassan, S.B. and Aigbodion, V.S., 2010. The study of the microstructure and interfacial reaction of Al-Cu-Mg/bagasse ash particulate composite. *Journal of Alloy and Compounds*, 491: 571–574.

Naresh, P., 2006. Development and characterization of metal matrix composite using red mud an industrial waste for wear resistant applications. *Ph.D. Thesis,* Department of Mechanical Engineering, National Institute of Technology Rourkela (India) January, pp. 23–34.

Chapter 11

Development of Polymer Ceramics with Post Consumer Polyethylene Terephthalate Waste Bottles

E.A. Olugbade[1]*, O.K. Badara*[2] *and K.F.K. Oyedeko*[3]

Department of Chemical and Polymer Engineering, Lagos State University, Lagos, Nigeria

ABSTRACT

Polyethylene Terephthalate (PET) post-consumer waste bottles and mineral aggregates were utilized as components of composite material. The study of the effect of the variation of the components: PET, cement, and sand in the composition of the formulated polymer ceramics on properties such as compressive strength, water absorption and heat treatment were conducted. Thermal treatment of the composite materials products from: sand and PET; and sand, cement and PET mix used; influenced the physical and mechanical properties of the composite materials samples obtained. These components mix under efficient temperature controlled condition and in appropriate proportions yielded materials in the PET and sand system; PET, sand and cement system, with compressive strength of 14.4MPa, 10.16MPa and percentage water absorption of 1.09 and 0.9 respectively.

Keywords: *Polyethylene Terephthalate (PET) waste bottles, Sand, Cement, Heat treatment, Compressive strength, Water absorption.*

E-mail: 1: ebenolugbade@yahoo.com, 2: olamidebadara@gmail.com, 3: kfkoyedeko@yahoo.com

Introduction

Polyethylene Terephthalate (PET) is a polymer in widespread use in the production of bottles and containers of water and carbonated drinks. Due to the high consumption of beverages and water, the production of PET bottles has greatly increased and has resulted in the increase of post-consumer PET waste. Polyethylene Terephthalate (PET) polymer is neither compostable nor environmentally biodegradable, and thus, suffers disposable problems. Since Polyethylene Terephthalate is resistant to microbial attack, PET bottles left to litter especially in habituated locations pose serious dangers not only to health but also to socio-economic well-being of humans as well. Recycling has become a viable alternative to the long term accumulation of the garbage. Bottle-grade PET is known to be of high quality among the various PET grades.

Post-consumer PET waste can be cleaned, shredded into flakes and recycled into Polyester Fibers, which serves as a base material for the production of clothing, pillows, carpets, chemically recycled or depolymerized and used for the production of bottles. (Badara, 2011).

Nevertheless, PET particularly following melting and re-solidifying of post-consumer recyclable waste can be quite brittle as such can be inappropriate in some applications.

Polymer ceramics studied under this investigation comprises of formulations of recyclable post-consumer Polyethylene Terephthalate flakes, sand and portland cement.

Significance of the Study

The utilization of the PET waste bottles as raw material for other production can help to reduce environmental pollution caused by improper disposal of these wastes. Polymer ceramics can serve as an alternative to purely cementitious concrete and other conventional materials probably due to superior material behaviour.

Aims and Objectives

The objective of the study was to utilize waste PET bottles as a component of composite materials and to study the effects of the variations in the composition of formulated polymer ceramics properties such as compressive strength, heat treatment, and water absorption.

The aim of the study focuses on the use of chemically unmodified post-consumer PET bottles, sand, and portland cement to produce polymer ceramics of superior materials behaviour.

Cement: The chemical composition of Portland cement is generally presented as shown bellow in Table 11.1.

Sand: is a naturally occurring granular material composed of fine divided rock and mineral particles. The composition of sand is highly variable, depending on the local rock source and conditions but most common constituent of sand is silica (Badara, 2011).

Table 11.1: Components of Portland Cement

Components	*Mass Content (per cent)*
CaO	58–66
SiO_2	18–26
Al_2O_3	4–12
Fe_2O_3 + FeO	1–6
MgO	1 -3
SO_3	0.5–2.5
K_2O and Na_2O	1

Adapted from: Toughened Polyethylene Terephthalate, United States Patent 7351756

Polyethylene Terephthalate PET: is a crystalline polymer that exhibits high modulus, high strength, high melting point, good electrical properties, and moisture and solvent resistance. Primary applications of PET include blow-moulded beverage bottles, fibres for wash-and-wear, wrinkle-resistant fabrics, and films that are used in food packaging; electrical applications (capacitors, etc), magnetic recording tape, and graphi carts.

Method

Raw Materials Used

1. Post-consumer PET bottles
2. Portland cement
3. Sharp sand (Diameter less than or equal to 0.6mm)

Procedures

Post-consumer PET bottles, about five hundred pieces, 75cl each were collected and washed with mild detergent and warm water to get rid of impurities and food particles. The covers and labels were removed. The cleaned PET bottles were crushed to manageable sizes to give PET flakes.

10-90gm weights of PET flakes in accordance with formulated measure were weighed and melted at a temperature of about 250°C. Formulated measures of Portland cement; sand; cement and sand were weighed and added to proportionate PET melt (Table 11.2, for formulation weight ratio). This was quickly stirred, and the resulting mixture was returned to the furnace for further heating for about 30seconds to 5minutes until homogeneous melt was obtained. The resulting mixture was then quickly poured into a pre-heated mold and covered. Sample was allowed to cool, solidify and then removed from the mold.

Heat Treatment

Samples were heated at 100°C for twenty four hours, another set of samples were also heated at 160°C for twenty four hours for compressive strength test.

Compressive Strength Test

Compressive strength test was carried out on the samples at the Lagos State Materials Testing Laboratory using standard equipment.

Water Absorption Test

For the water absorption test, samples were submerged in water for seven days, then dried in air and weighed. Samples were then placed in an oven at a temperature of 120pC for about three hours, allowed to cool and weighed.

Results and Discussion

Results

Characteristics of polymer ceramics composite samples obtained are depicted in Table 11.2 showing the sample composite weight ratio, compressive strength and per cent water absorption of the samples submerged in water for 7 days, and the compressive strength of samples heat treated at 100°C and 160°C for 24hrs respectively.

Table 11.2: Various Characteristics of Sand, PET, Cement and Sand, PET Composites

Sample	*Composite (Wt. Ratio)*	*Compressive Strength (MPa)*	*Per cent Water Absorption*	*Remarks*
A	20Sand7PET3Cement	6.6	0.9	Submerged in water for 7days
		9.32		Heat treated at 100°C for 24hrs.
		10.16		Heat treated at 160°C for 24hrs
B	20Sand8PET2Cement	6.50	1.08	Submerged in water for 7days
		11.16		Heat treated at 100°C for 24hrs
		12.12		Heat treated at 160°C for 24hrs
C	20Sand9PET1Cement	6.6	1.29	Submerged in water for 7days
		12.8		Heat treated at 100°C for 24hrs.
		14.28		Heat treated at 160°C for 24hrs.
D	3Sand2PET	9.32	1.56	Submerged in water for 7days
		12.8		Heat treated at 100°C for 24hrs.
		14.36		Heat treated at 160°C for 24hrs.
E	2Sand1PET	10.04	1.09	Submerged in water for 7days.
		12.84		Heat treated at 100°C for 24hrs.
		14.40		Heat treated at 160°C for 24hrs
F	1Sand1PET	6.19		
G	2Sand3PET	3.40		

Figure 11.1 shows the per cent water absorption of composite samples A, B, C containing sand, PET, cement and composite samples D and E containing sand and PET.

Figure 11.2 presents the compressive strength of composite samples A-F as prepared and cured in water for 7days, where each sample A, B, C, contains sand, PET, cement and composite samples D, E, F, contain sand, PET each.

Figures 11.3 and 11.4 shows the compressive strength of composite samples heat treated at different temperatues (100pC and 160pC respectively) where each composite sample A, B, C contains sand, PET, cement while composite samples D, E contain sand, PET each, see Table 2 for composites weight ratio.

Discussion

Characteristics of samples prepared from PET and cement samples with cement content between 25 per cent -60 per cent showed that samples became tougher as cement content increased but began to reduce in toughness as cement content was increased above 30 per cent. However, generally, samples appeared fragile and brittle, and due to observed cracks on the samples, compressive strength test and water absorption test were discarded. This finding appears to be in agreement with US Patent 7351756 granted to Simmons and Simmons (2008). Their experiment showed that it was difficult to add more than 15wt per cent cement to PET flakes because the cement appeared to separate from the flakes resulting in non-homogenous melt and injection moulded material. The composition of cement/PET composite in our investigation was 25 per cent - 60 per cent cement. Composite samples prepared from PET and sand with sand content of 40 per cent to about 67 per cent were investigated, so also samples prepared from PET, sand and cement were investigated. Various characteristics of these samples are shown in Table 11.2 and enunciated in Figures 11.1–11. 4.

Simmons and Simmon (2008) reported that US Pat.No. 6, 583,217B1 to Li *et al.,* is directed to composite material composed of fly-ash and waste PET. U. S. Pat. 7351756 to Simmons *et al.* (2008) is directed to a composite material composed of Porland cement about 5wt per cent - 50 per cent (wt) and waste PET. Simmons *et al.* (2008) stated that the reason that Portland cement toughens PET was not well understood. Three explanations were offered for observed toughening effect as follow:

First, the cement may toughen the polymer by inhibiting crack propagation. Second, the cement may act as a getter for water which is known to greatly reduce PET chain length. Third, the alkali metal in the cement may act as a weak cross-linking agent forming the larger chain length and thus resulting in greater toughness.

Figure 11.1 appears to show the effect of decreasing cement content on per cent water absorption on composite weight ratio (Table 11.2) with sample A with the highest cement content being the most water resistant. Sample E with a weight ratio of sand:PET 2:1 appears to be more water resistant more than Sample D with a weight ratio of sand:PET 3:2.

Further more from our investigations, Table 11.2 and Figures 11.3 and 11.4, indicate that heat treatment effectively influenced the compressive strength of composite samples. In glass-ceramics, thermal treatment nucleation at lower temperature and heating at high temperature has been found to produce glass ceramics with grains (Buh and Olugbade, 1978). It is well known in ceramics that

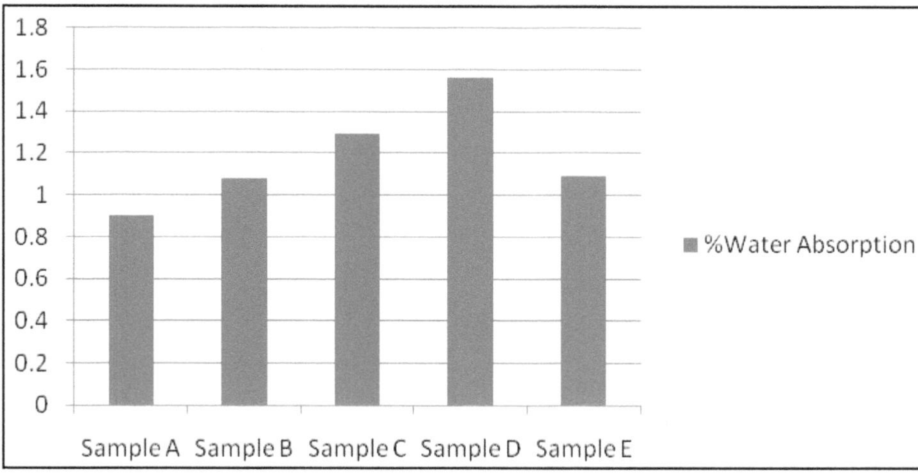

Figure 11.1: Water Absorption (per cent) of Samples

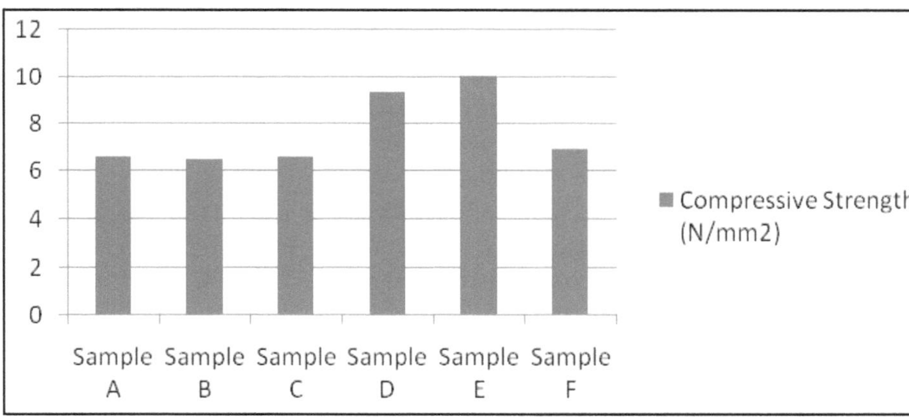

Figure 11.2: Compressive Strength of as Prepared Samples Cured in Water for 7 Days

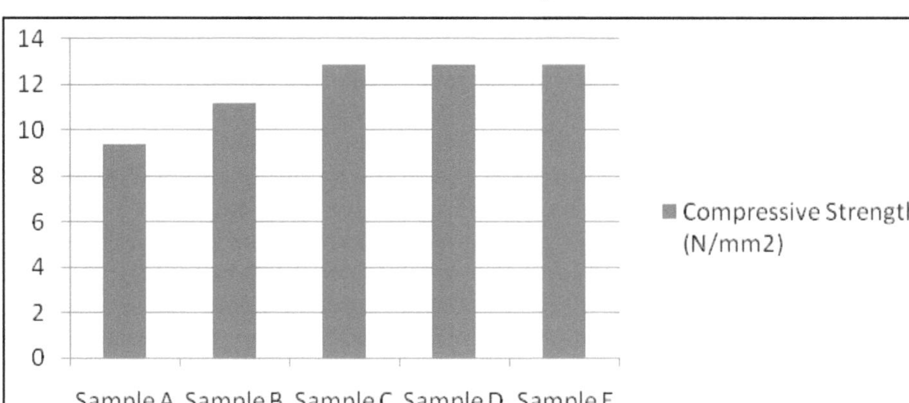

Figure 11.3: Compressive Strength of Samples Heated at 100°C for 24 hrs

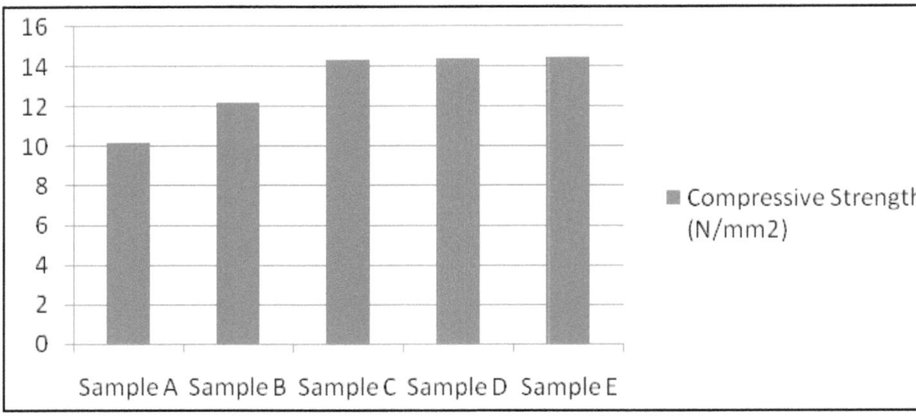

Figure 11.4: Compressive Strength of Samples Heated at 160°C for 24 hrs

mechanical properties of materials are greatly enhanced with grains development from the liquid phase (Kingery *et al.,* 1978). Thus, it could be inferred that increase in compressive strength of thermally treated samples of composites (see under Remark on Table 11.2) in PET matrix is greatly influenced by grains development in the PET composites system of our investigation. Compressive strength increases with increase in heat treatment temperature.

The composite materials are amenable to various uses. Thus, custom made composite material for various uses could be obtained in polymer-ceramics system of material aggregates with PET waste bottles Figures 11.2–11.4. The compostites materials obtained could be used where prevention of water absorption is essential *e.g.* water-proof vessels, drainage channels and swimming pools in contrast to cement concrete blocks which have water absorption greater than 15 per cent. The composite material obtained may be used as in masonry concrete or low to moderate strength concrete since the compressive strength of the material falls between 6.5–14.5MPa. Composite material obtained could also be considered for use for decorative purposes because of its glossy appearance and sprayable surface. It could probably be used for production of decorative wall tiles for interior purposes, roof and floor tiles, and for other decorative purposes as it compares favourably with data on water absorption and compressive strength of tiles (Table 11.3). From Table 11.3 it could be derived that composite materials obtained in this investigation could be used in the manufacture of glazed wall-tiles for interior purposes, roof and floor tiles of superior quality with respect to their ability for water tightness and relative high compressive strength of potential products. The values of per cent water absorption of the samples are well below the minimum value of 18 per cent water absorption requirered for glazed wall-tiles for interior purposes and 7 per cent water absorption for floor tiles. Similarly, the compressive strengths of the composite materials products are well above minimum required values for the glazed wall tiles, roof and floor tiles (Table 11.3).

Table 11.3: Water Absorption and Compressive Strength Requirement of Tiles; and Per cent Water Absorption and Compressive Strength of Composite Samples as Prepared and Cured in Water for 7 Days in this Investigation

Material	*Water Absorption*	*Compressive Strength*	
#Glazed wall-tiles for Interior Purposes	*Not more than 18 per cent*	*Not less than 90psi*	
#Floor tiles	*Not more than 7 per cent*	*Not less than 250psi*	
A 20Sand7PET3Cement	0.9	MPa	Psi
		6.60	957.25
B 20Sand8PET2Cement	1.08	6.50	942.75
C 20Sand9PET1Cement	1.29	6.60	957.25
D 3Sand2PET	1.56	9.32	1351.75
E 2Sand1PET	1.09	10.04	1458.18

#: Ceramics Tiles Institute of America [online], 2011.

Conclusion

Waste PET bottles when properly cleaned, flaked, melted and thoroughly mixed with sand under efficient temperature controlled conditions and in the right proportion yeided a composite material with compressive strength of 14.4Mpa (2088.54psi) and percentage water absorption of 1.09. Thermal treatment, proportion of sand and PET; cement, sand and PET used influenced the physical and mechanical properties of the composite polymer ceramics samples obtained.

The composite ceramic materials obtained could be used in the manufacture of glazed wall-tiles for inerior purposes, floor and roof tiles, construction of roof tops, water-proof vessels and drainage channels.

References

Badara, O.K., 2011. Formulation and analysis of polymer concrete prepared from waste polyethylene terephthalate bottles. *B.Sc. Project Report*, Dept. of Chemical and Polymer Engineering, Lagos State University, Lagos, Nigeria.

Buh, M. and Olugbade, E.A., 1978. Semiconducting glass-ceramics on the base of Li_2O-Al_2O_3-SiO_2 System with added TiO_2. In: *Contemporary Inorganic Materials*, (Eds.) G. Pettzow, and W.J. Huppman. Stuttgart: Dr. Riederer Verlag GmbH, pp. 131–138.

Ceramics Tile Institute of America [online], http//www.ctia.org/report/fr 41 html [accessed 17th January, 2011].

http//www.freepatentsonline.com/7351756.html[accessed 8th of August, 2010].

Kingery, W.D., Broven, H.K. and Uhlmann, D.R., 1978. *Introduction to Ceramics*, 2nd Edn. John Wiley and Sons.

Simmons, Walter John and Simmons, Walter Neal, 2008. *Toughened Polyethylene Terephthalate*. U.S. Pat. 7351756.

Chapter 12

Water Melon Peels Board: Alternatives to Wood-based Particleboard Composites

U.D. Idris[1], *V.S. Aigbodion*[2]*, *C.U. Atuanya*[3] *and Abdullahi*[1]

[1]*Samaru College of Agriculture, Ahmadu Bello University, Zaria, Nigeria*
[2]*Department of Metallurgical and Materials Engineering, Ahmadu Bello University, Samaru, Zaria, Nigeria*
[3]*Department of Metallurgical and Materials Engineering, Nnamdi Azikiwe University, Awka, Nigeria*

ABSTRACT

The aim of this study was to investigate the suitability of using water melon peels as alternatives to wood-based particleboard composites. The water melon peels composite boards were produced by compressive moulding using recycled low density polyethylene (RLDPE) as a binder. The RLDPE was varies from 30 to 70wt per cent with interval of 10wt per cent. The microstructure, water absorption(WA), thickness swelling index(TS), modulus of rupture (MOR), modulus of elasticity (MOE), internal bonding strength(IB), impact strength and wear properties of the boards were determined. The results showed that high modulus of rupture of 11.45N/mm^2, MOE of 1678N/mm^2, IB of 0.58N/mm^2, wear rate of 0.31g were obtained from particleboard produced at 60wt per cent RLDPE. The uniform distribution of the water melon particles and the RLDPE in the microstructure of the composites board is the major factor responsible for the improvement in the mechanical properties. The results showed that the MOE, MOR and IB meet the minimum requirements of the European standards, for general purpose like panelling, ceiling, partitioning. Hence, water melon particles can be used as a substitute to wood-based particleboard for general purpose

* *Corresponding Author:* E-mail: aigbodionv@yahoo.com

applications also, besides being environmental friendly by using watermelon and RLDPE in production of particleboard, this alternative to wood-based particleboard is very cost-effective.

Keywords: *Board composites, Water melon particles, Mechanical, Physical and Wear properties.*

Introduction

The demand for wood composites from waste wood has been increasing as timber resources in natural forests decline. The use of renewable biomass as a raw material in composites production was one approach and the use of renewable biomass may result in several benefits such as environmental and socio-economic [1]. Today, renewable biomass are mostly accepted as waste materials and are mostly ploughed into the soil or burnt in the field.

Traditionally, particleboard has been made out of wood-based fibers bound together using a formaldehyde resin. The desired thickness is achieved by using a hot press that forms the board into sheets [2]. Particleboard has a homogenous structure and can be manufactured in different sizes, thickness, densities and grades for numerous uses, making it a desirable material with which to work [2].

One of the major challenges associated with wood-based particleboard is the use of Formaldehyde resin Formaldehyde is a volatile, colorless gas with a strong odor that is commonly used in industrial processes, particularly in manufacturing building materials. Pressed wood products, such as wood-based particleboard and medium density fiberboard, are made using adhesive resins containing urea-formaldehyde. Off-gassing levels are at their highest when the products are new, with emissions tapering off as they age. Exposure to formaldehyde in concentrations greater than 0.1 parts per million (ppm) can cause nasal and throat congestions, burning eyes, or headaches as well as increasing the risk of developing cancer[2].

According to the end uses of wood-wastes and their possible reuse products, particleboard has found typical applications as flooring, wall and ceiling panels, office dividers, bulletin boards, furniture, cabinets, counter tops, and desk tops[3, 4], and it seems that the manufacture of particleboard from biomass wastes is the most common way to reuse such waste materials[1, 6].

Wood composite industries demand more wood raw material everyday despite the fact that the forest resources are diminishing. The decline in wood material source has led researchers to study non-wood ligno-cellulosic biomass utilization in composite manufacturing including particleboard. Agricultural waste materials and annual plant fiber have become alternative raw materials for the production of particle or fiber composite materials. The most frequently referred alternative non-wood materials are flax, bagasse, hemp, reed, and cereal straws such as rice and wheat straw [1]. Today, chemical pulp and panel products using wheat straw and other crop residues are being commercially manufactured in a number of countries including Nigeria [6]. There is still, a growing need to find alternative sources of raw materials for composite manufacturing. Therefore, the aim of this study is to investigate the potential utilization of water melon peels using recycled low density polyethylene as binder in particleboard production as a supplement, to alleviate the shortage of raw material in forest industry and also to avoid the toxic nature of Formaldehyde commonly used binder.

Materials/Equipment

The water melon peels used in this work was obtained from 'Sabon Gari' area of Zaria in Kaduna state Nigeria (Figure 12.1)

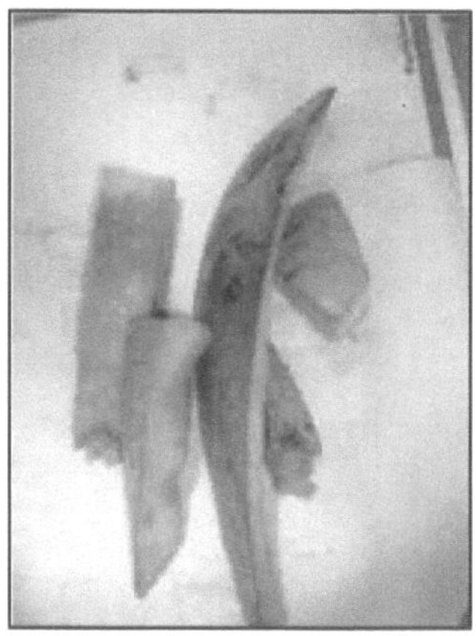

Figure 12.1: Photograph of the Water Melon Peels

Waste Pure water sachet (RLDPE)" used was collected literally from the streets of Zaria and around refuse dumps. Cut in to small pieces with the aid of blades and pair of scissors (Figure 12.2).

Figure 12.2: The RLDPE Resin Used

Equipment used in this research are-, Metal mould, hydraulic press, Avery Denison impact tester, Instron machine, grinding and polishing machine, Scanning electron microscope(SEM) and Pin on Disc wear machine.

Method

The particle size analysis of the water melon peels particles was carried out in accordance with BS1377:1990[7]. About 100g of the particles was placed onto a set of sieves arranged in descending order of fineness and shaken for 15minutes which is the recommended time to achieve complete classification, the particle that was retained in the BS. 100μm was used in this study (Figure 12.3).

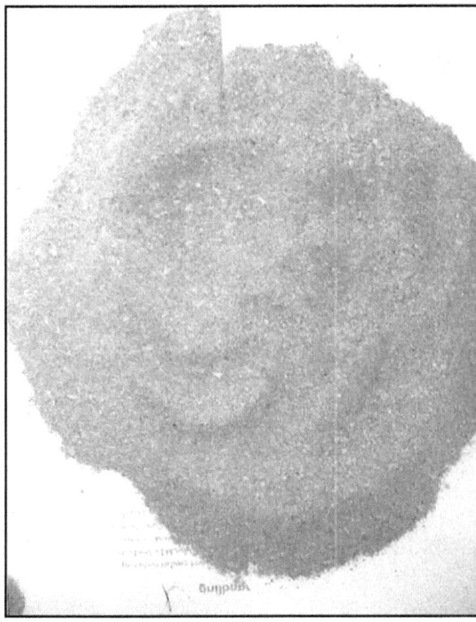

Figure 12.3: Photograph of the Water Melon Peels Sieve Particles

Metal Molds was used in the production of the board composite samples. Each mold had a cavity to accommodate the board composite samples. The dimensions and shapes of cavities were made according to the size and shape of the samples as per ASTM Standard D 638-90 for tensile testing and ASTM Standard D 790-97 for flexural testing[7]. After drying the water melon particles and RLDPE in oven at 105°C, the agro-waste particles and the RLDPE was then compounded in a two roll mill at a temperature of 130°C, into a homogenous mixture. Board production was carried out on an electrical heated hydraulic press. The mixtures were then placed in a rectangular mould. The boards were pressed by varying the RLDPE Resin from 30-70wt per cent with 10wt per cent interval. At the end of press cycle the board was removed from the press for cooling.

The scanning electron microscope (SEM) JEOL JSM-6480LV was used to identify the surface morphology of the board composite samples. The surfaces of the board composite specimens are examined directly by scanning electron microscope JEOL JSM-6480LV. The samples were washed, cleaned thoroughly, air-dried and are coated with 100 Å thick platinum in JEOL sputter ion coater and observed SEM at 20 kV. The digitized images were recorded.

The basic method of determining the density of board composite samples was by measuring the mass and volume of the sample used. A clean sample is weighed accurately in air using a laboratory balance and then suspended in water. The weight of the sample when suspended in water was determined, the volume of the sample was determined from the effect of displacement by water (Archimedean principle). The density of the sample was then estimated from equation below [8]:

$$Density = \frac{Mass}{Volume}$$

Specimens with dimensions of 50 mm x 50 mm were prepared for evaluation of the thickness swelling. The thickness at the middle of the test specimen was measured with a micrometer. Then the test specimens were place into water in parallel for 30 mm and soaked for 24 h before further measurement of the thickness. The Thickness swelling rate (TS) was determined from the following formula [8].

$$TS_{24} = \frac{(t_{24} - t_0)}{t_0} \times 100$$

where, TS is the thickness swelling rate (per cent), t_0 and t_{24} are the thickness at the middle of the test specimen. The values of the WA as percentages were calculated [8]

$$WA(t) = \frac{W(t) - W_0}{W_0} \times 100$$

where, WA(t) is the water absorption (per cent) at time t, W_0 is the initial weight, and W(t) is the weight of the sample at a given immersion time t.

Bending specimens of 50 mm wide 275 mm long were cut from each full particleboard. A concentrated bending load was applied at the center with a span of 15 times the thickness of the specimen. The bending modulus of elasticity (MOE) and modulus of rupture (MOR) were calculated from load deflection curves according to the following formula [8]:

$$MOR = \frac{3P_b L}{2bh^2}$$

$$MOE = \frac{P_{bp} L^3}{42bh^3 Y_p}$$

where, P_b is the maximum load (N), P_{bp} is the load at the proportional limit (N), Y_p is the deflection corresponding to P_{bp} (mm), b is the width of the specimen (mm), h is the thickness of the specimen (mm), and L is the span (mm).

The tensile strength perpendicular to the surface was determined using three conditioned specimens of 50 mm x 50mm from each particleboard. The rupture load (*Ps*) was determined and internal bond strength (IB) was calculated using the following formula[7].

$$IB = \frac{Ps}{bl}$$

where, *Ps* is the rupture load, and *l* is the length of the specimen.

The impact test of the board composites sample was conducted using a fully instrumented Avery Denison test machine. Charpy impact tests was conducted on notched samples. Standard square impact test sample of measuring 75 x 10 x 10 mm with notch depth of 2 mm and a notch tip radius of 0. 02 mm at angle of 45°was used [7].

A pin-on-disc test apparatus was used to investigate the dry sliding wear characteristics of the composites as per ASTM G99- 95 standards [9]. The wear test was conducted using sliding speed of 1m/s, applied load of 10N and sliding distance of 1000m.

Results and Discussion

Macrostructural studies of the particleboard revealed a uniform distribution of agro-waste particles with the RLDPE binder. The distribution of particles is influenced by the compounding of the particle and the binder which resulted in good interfacial bonding (Figure 12.3). However, during the production of the boards, RLDPE resin binder at 10-20wt per cent addition did not give good bonding, as the RLDPE binder increased beyond 20wt per cent good bonding of the particles and the resin was obtained. This observation is in par with the earlier studying of Atuanya *et al.* [2].

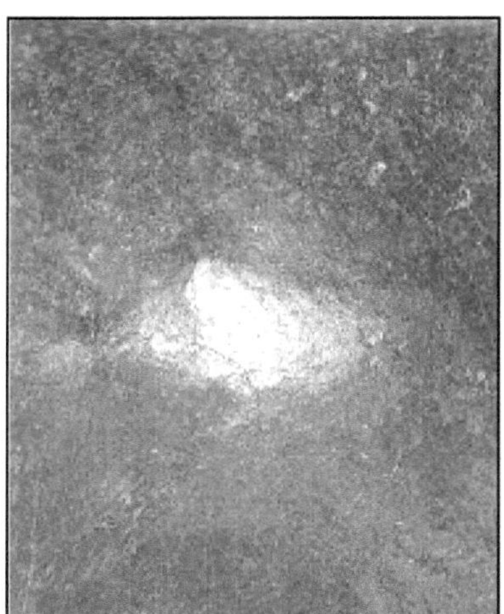

Figure 12.3: Photograph of the Developed Particle Board

The morphologies of the particleboard composites by SEM are shown in Figures 12.4–12.6. Morphological analysis using SEM clearly shows difference in the morphology of the particleboard composites (Figures 12.4–12.6). The microstructure clearly shows that when the agro-waste particle was added to the RLDPE resin morphological change in the structure takes place.

Figure 12.4: SEM Microstructure of the Water Melon Peels Particleboard with 30wt per cent RLDPE Resin

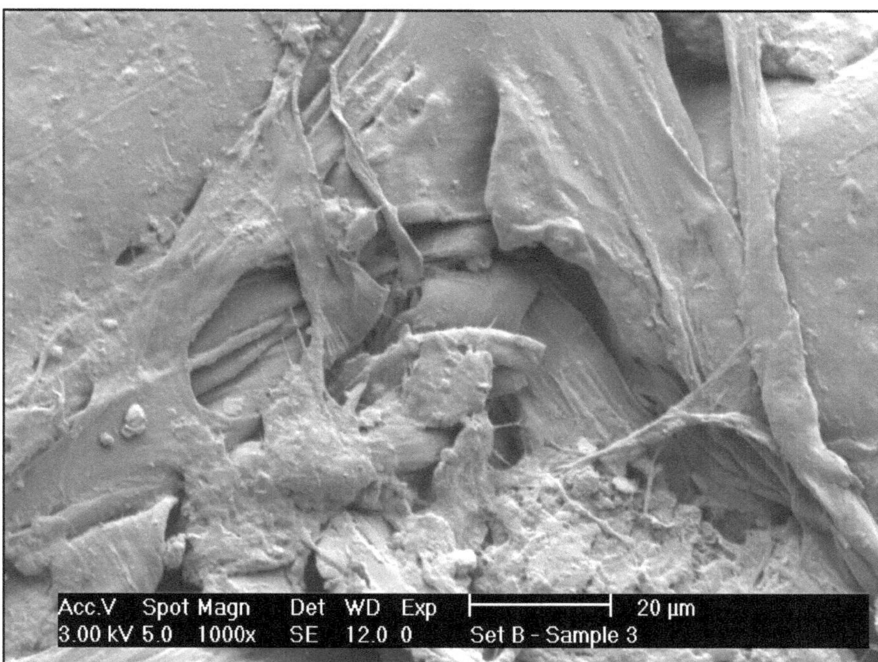

Figure 12.5: SEM Microstructure of the Water Melon Peels Particleboard with 50wt per cent RLDPE Resin

Figure 12.6: SEM Microstructure of the Water Melon Peels Particleboard with 70wt per cent RLDPE Resin

The microstructure reveals that there are small discontinuities and a reasonably uniform distribution of water melon particles and the RLDPE resin. The water melon particles phase is shown as white phase, while the resin phase is dark [9]. The agro-waste particles are embedded within the amorphous matrix composed of randomly distributed in the matrix planar boundaries. The surface of the agro-waste particles is smooth indicating that the compatibility between particles and the resin was good. It can be seen that the agro-waste particles are not detached from the RLDPE surface as the weight fraction of agro-waste particles increased in the resin, this is due to good interfacial bonding between the resin and the particles [2, 8]. This good bonding was achieved from the compounding of the agro-waste particles and the resin.

The density, thickness swelling (TS) and water absorption (WA) of the particleboards are shown in Figures 12.7–12.9. The values obtained for the thickness swelling (TS) and water absorption (WA) of the particleboards were high (Figures 12.7–12.8). This is due to, not using water repellent agents in the particleboard manufacturing. But these results are better than that resulted by other authors; which were obtained for wood based particleboard [6, 11-13].

The water melon particles affected the WA and TS properties negatively. Similar results were found by Ntalos and Grigoriou [11]. Decreasing the resin, increased the WA and TS of particleboards. This may be due to high solubility values of the particles.

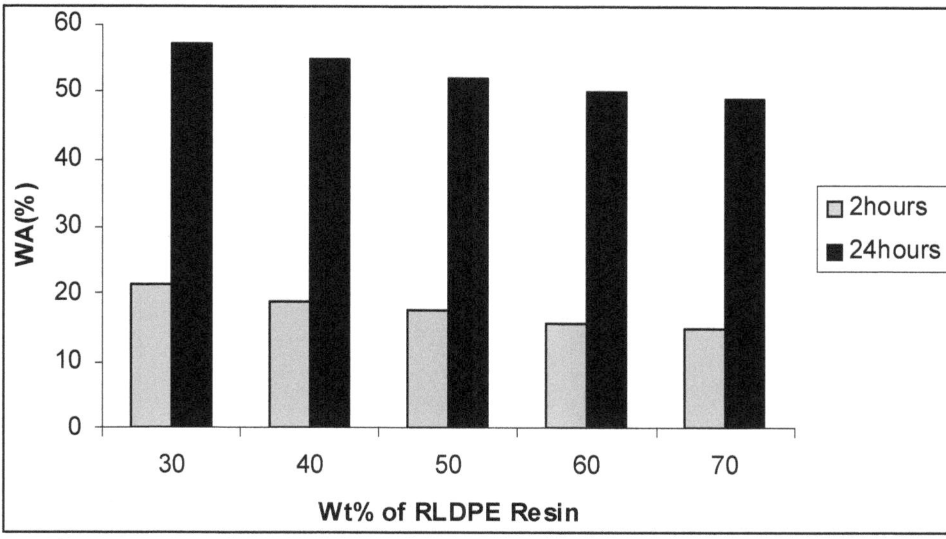

Figure 12.7: Variation of Water Absorption with RLDPE Addition

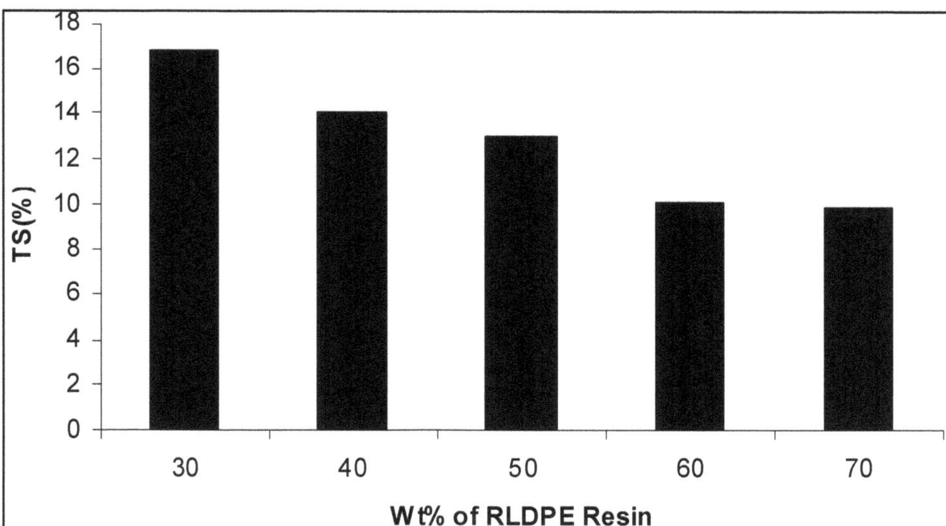

Figure 12.8: Variation of Thickness Swelling with RLDPE Addition

In addition, wall thickness of particles was found to be within the ranges of that of common wood species [9].

The density profile of a board is dependent on the particle configuration, moisture distribution in the mat, hot press temperature and rate of closing, resin reactivity and the compressive strength of the particles [9]. The boards density increased with increasing the weight fraction of the RLDPE resin (Figure 12.9). Increasing of board density from 0.50 to 0.70 g/cm^3 decreased the TS and WA for 2 h immersion. This is

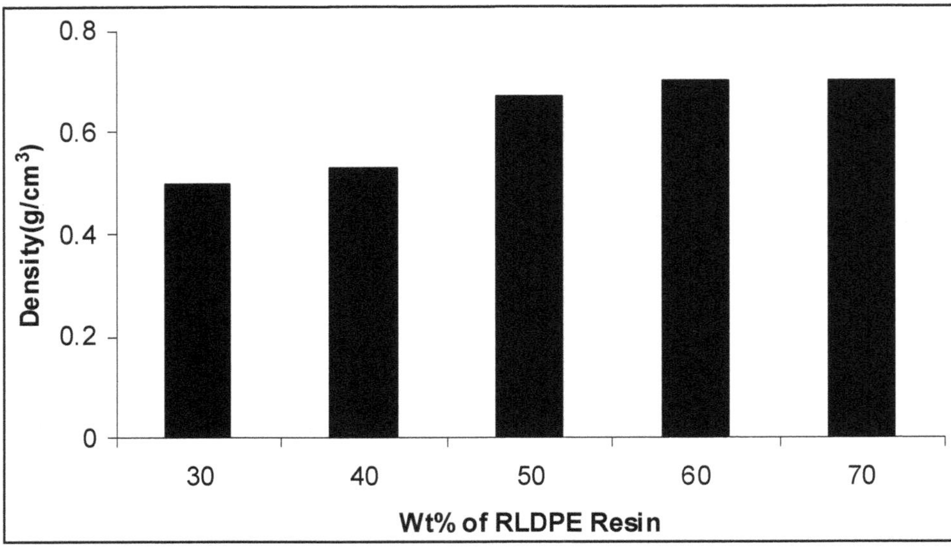

Figure 12.9: Variation of Density with RLDPE Addition

due to low porosity and difficult diffusion on the high board density. The swelling that occurs is the sum of two components, namely, swelling by hygroscopic particles and the release of compression stresses imparted to the board during the pressing of mat in the hot press [5]. The release of compression stresses, known as springback, is not recovered when the board is in a dry state

The values of modulus of elasticity (MOE), modulus of Rupture (MOR), internal bond strength (IB) strength and impact strength (IM) of the boards are shown in Figures 12.10–12.12.

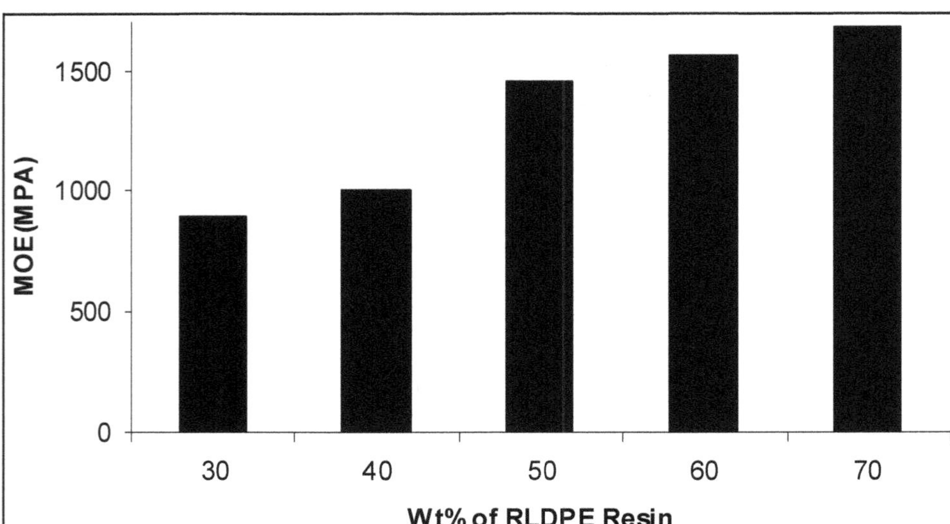

Figure 12.10: Variation of Modulus of Elasticity with RLDPE Addition

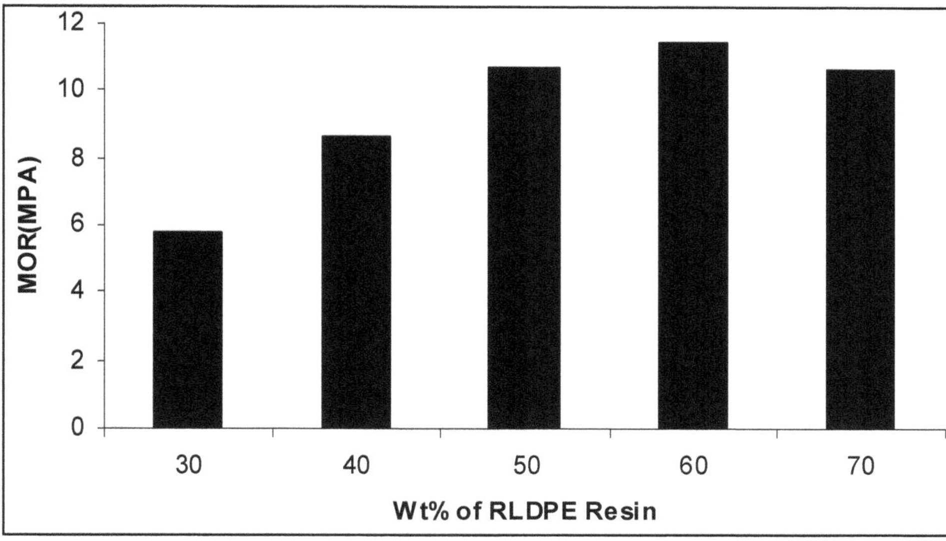

Figure 12.11: Variation of Modulus of Rupture with RLDPE Addition

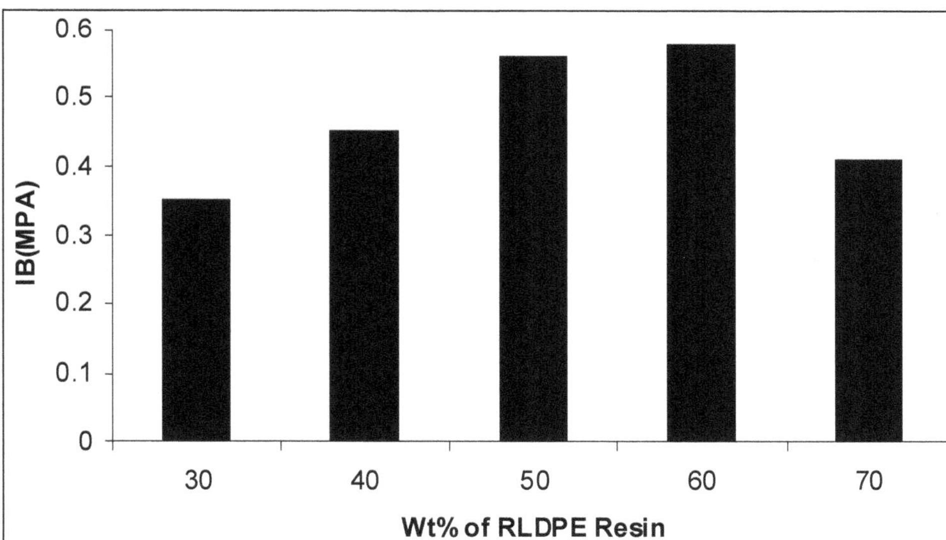

Figure 12.12: Variation of Internal Bonding Strength with RLDPE Addition

Tensile modulus and internal bonding strength measurements are amongst the most important indicators of strength in a material and are most widely specified property. Tensile modulus, an indication of the relative stiffness of a material [11-14]. Tensile test is a measurement of the property of a material to withstand forces that tend to pull it apart and to determine to what extent the material stretches before breaking.

The improvement in internal bonding strength and tensile modulus was noticed with the developed particle board. This indicates that use of water melon particle and RLDPE in production of the particleboard improves the load bearing capacity of the board. Similar observations have been reported by [2, 8, 10-11] for other particleboard. In addition, the developed particleboard deforms less until maximum load, which gives a higher tensile modulus. The increase in Young's modulus with increasing RLDPE is expected since the addition of RLDPE to water melon particles increases the stiffness of the particleboard. The presence of polar group in the RLDPE may contribute to electrostatic adsorption between RLDPE and water melon particles. This phenomenon is driven by different charges acting on RLDPE or water melon particles surfaces; this mechanism will strengthen the RLDPE/water melon particles interface. It will hold them together and increase their resistance to deformation [9], this helps in increment of the board modulus. The fairly uniform of water melon particles distribution has efficiently hindered the chains movement during deformation. This mechanism will increase the stiffness of the board as well as its tensile modulus.

The MOR ranged from 5.78 to 11.6N/mm^2 (Figure 12.11). The MOR requirements of 11.5 N/mm2 for general purpose boards by EN 312-2[12]. Particleboards made from 30wt per cent RLDPE resin had MOR lower than the requirement for general purpose. In addition, although increasing of RLDPE Resin addition increasing the MOR upto 60wt per cent, beyond this level no further increased in MOR was obtain.

The range of data in IB was from 0.35 to 0.58 N/mm^2 (Figure 12.12). The IB requirements of 0.24 N/mm2 for general purpose boards, 0.35 N/mm2 for interior fitments, load-bearing boards and 0.50 N/mm2 for heavy duty load bearing boards by [10-13] respectively.

The internal bond strength is comparable with values reported by Razali and Kuo [16] and Chew *et al.* [4]. The water melon peels board surpassed the mechanical strength requirements for general purpose applications specified by European standard. All of the particleboards produced, are within the recommended standard i.e for general purpose, interior fitments, load-bearing boards and heavy-duty load bearing boards.

The results of the impact strength shows that the impact strength of the board composites slightly increased with increase in RLDPE resin addition (Figure 12.13).

High strain rates or impact loads may be expected in many engineering applications of particleboard composite materials. The suitability of a particleboard composite for such applications should therefore be determined not only by usual design parameters, but by its impact or energy absorbing. The steep increase in the impact strength board composites could be attributed to the presence of particles well bonded by the RLDPE resin; this factor leads to increase in impact energy.

The results obtained for the wear rate of the developed particleboard are shown in Figure 12.14, from the results there is an appreciable reduction in the wear rate of the particleboard with the addition of the RLDPE upto 60wt per cent for example: wear rate obtained are 0.56g at 30wt per cent and then to 0.31g at 60wt per cent RLDPE. The decrease in wear rate of the composites could be attributed to higher load bearing capacity of hard particleboard and better interfacial bond between the particle and the matrix reducing the possibility of particle pull out which may result

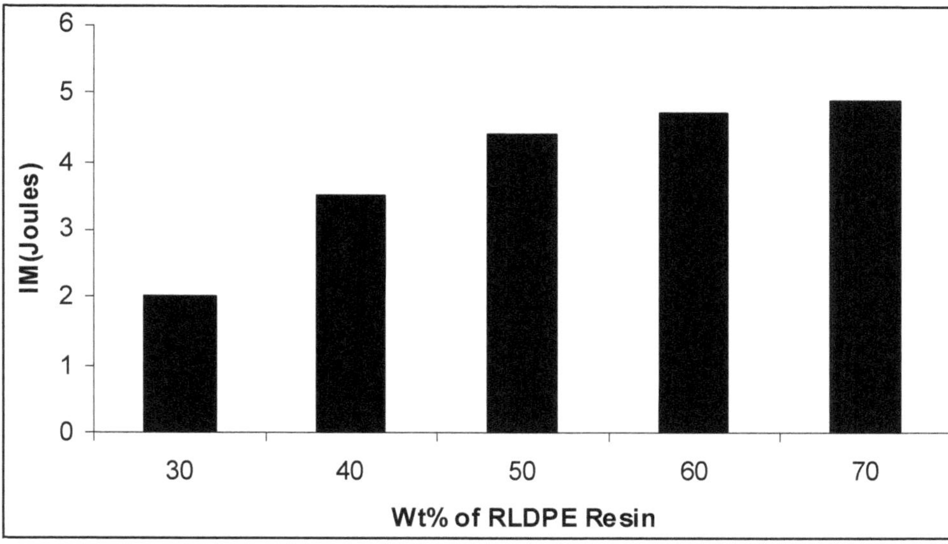

Figure 12.13: Variation of Impact Strength with RLDPE Addition

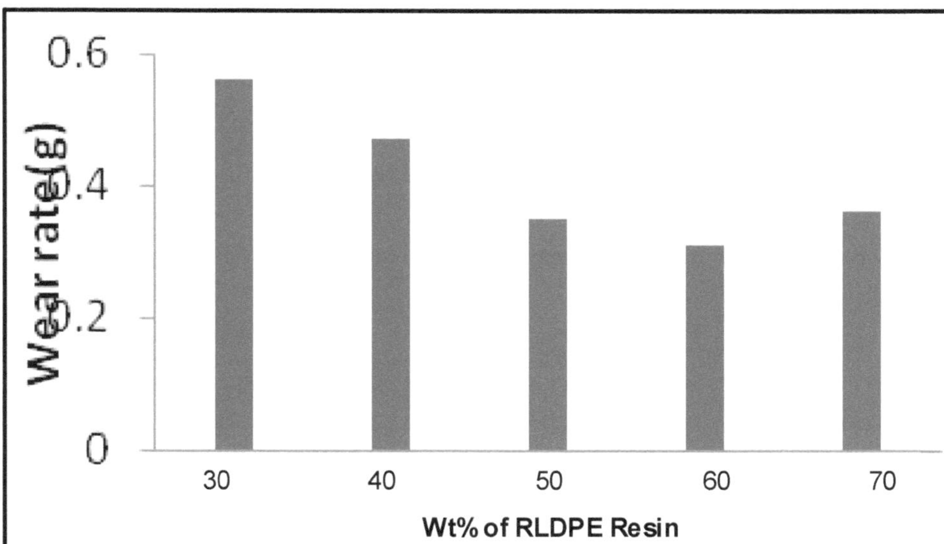

Figure 12.14: Variation of Wear Rate with RLDPE Addition

in higher wear (Figure 12.14) [9]. The increase in wear rate beyond 60wt per cent RLDPE may be attributed to low values of internal bonding strength obtained (Figure 12.12). The wear mechanism reported was adhesion and delamination. This observation is on par with the one observed by Aigbodion *et al.* [9]. It is well known that wear process involve fracture, tribochemical effects and plastic flow. Transitions between regions dominated by each of these, commonly give rise to changes in wear rate with load.

The developed particleboard exhibited the higher wear resistance under the sliding distances. This behaviour can be attributed to the presence of watermelon particles on the counter surface, which act as a transfer layer and effective barriers to prevent large-scale fragmentation of RLDPE. This is also evident from the small size of the wear debris observed. The wear resistance of the particleboard is improved by preventing direct contacts that induce subsurface deformation. The addition of hard watermelon particles improves the resistance to seizure. The water melon particulate allows considerable thermal softening effects without having adverse effects on the wear behaviour [9]. The watermelon particles also cause higher hardness and less coefficient of thermal expansion within the RLDPE resin. The presence of the ceramic particles provides a higher thermal stability, increased abrasion and sliding wear resistance at high and delays the transition from mild to severe wear[9].

The particleboard produced with 60wt per cent RLDPE and watermelon particles were compared with that of Flax-board made with flax fibers and urea resin glue (Urea Formaldehyde) and that of Medium Density Fiber-board (MDF) made of wood composite materials(Table 12.1), which was tested under similar conditions. The results are in close agreement. This study showed that watermelon particles can work well with RLDPE when used in the production of the particleboard. Also, different components of RLDPE had significant influence on mechanical properties of particle board. Generally, the boards produced with the watermelon particles at 50-60 per cent RLDPE resin addition conform to the BIS specifications of density particleboards for general purpose requirements like panelling, ceiling, partitioning etc. (interior decoration). Besides being environmental friendly by using watermelon and RLDPE in production of particleboard, this alternative to wood-based particleboard is very cost-effective.

Table 12.1: Summary of the result findings compared with existing ones.

Properties	*MDF (www. finsa.es) [17]*	*MDF (www. spanogroup.be) [18]*	*Flax-board (www. linex-pg.nl) [19]*	*Present Study Watermelon particle at 60 wt per cent RLDPE*
Board density (g/cm^2)	0.85	0.60-0.65	0.35-0.8	0.75
Internal boding strength (N/mm^2)	0.90	0.40-0.45	0.1	0.65
Modulus of elasticity (N/mm^2)	NA	1400-1700	750	1650
Modulus of Rupture (N/mm^2)	NA	NA	NA	12.0
Bending strength (N/mm^2)	38	14.0-20.0	45	
Thickness swelling (per cent)	15	11-20	NA	10.05
Wear rate(g)	NA	NA	NA	0.31

Note: Flax-board is a particle-board composite of flax fibers and urea resin glue (Urea Formaldehyde).

Medium Density Fiber-board (MDF) is a wood composite material made by wood fibers.

NA: Not available.

Conclusions

This present research is centred on the development and characterization of the microstructure and properties of particleboard composites using water melon peels and RLDPE as a binder and as an alternative to wood and Urea Formaldehyde particleboard. From the above results and discussion, the following conclusions are made:

1. This work shows the successful fabrication of water melon peels and RLDPE resin particleboard composites by simple compressive moulding technique.
2. The density increased as the percentage of resin increases in the particles.
3. The percentage of thickness swelling and water absorption increased in decreasing the weight fraction of the RLDPE resin
4. The tensile properties obtained are in agreement with the results obtained from the analysis of impact strength.
5. The uniform distribution of the particles and the resin in the microstructure of the board composites is the major factor responsible for the improvement in the mechanical properties.
6. The wear rate decreased as the weight percentage of RLDPE increased upto 60wt per cent
7. The developed particleboard composites can be used in density particleboards for general purpose requirements like panelling, ceiling, partitioning etc. Since the properties of particleboard composites used in this area are compared favourably with the properties of the developed board composites at 50-60wt per cent RLDPE resin.
8. It is recommended that besides being environmental friendly by using watermelon and RLDPE in production of particleboard, this alternative to wood-based particleboard is very cost-effective.

References

1. Yamashita, Y., Watanabe, H. and Takeda, S., 1999. Rubberwood characteristics of its supply and development of its utilization. *Forest Research,* 71: 65–70.

2. Atuanya, C.U., Ibhadode, A.O.A. and Igboanugo, A.C., 2011. Potential of using recycled low-density polyethylene in wood composites board. *Tribology in Industry,* 33(1): 11–17.

3. MacCleery, R., 1995. Resiliency the trade mark of American forests. *Forest Prod. J.* 45: 19/28.

4. Chew, L.T., Nurulhuda, M.N., Ong, C.L. and Rahim, S., 1991. Particleboard from some plantation species. In: *Proceedings of a Regional Symposium on Recent Developments in Tree Plantations of Humid/Subhumid Tropics of Asia,* (Eds.) S.A. Abod *et al.,* 5–9 June 1989, Serdang. Universiti Putra Malaysia, Serdang, pp. 708–724.

5. Dalen, H., 1999. Factors to take into consideration when producing particleboard from straw. *Proceedings of the Meeting of the Eastern Canadian Section of the Forest Products Society*, p. 456–462.

6. Hall, H.J., Gertjajensen, R.O., Schmidt, E.L., Carl, C.G. and DeGroot, R.C., 1984. Preservative effect on mechanical and thickness swelling properties of Aspen wafer board. In: Proceeding of a Workshop on Durability of Str. Panels. Pencacola, Florida, USA, 234–247

7. ASTM E290, 1990. *Physical Testing Standards and Mechanical Testing Standards*, p. 1234–1345.

8. BSI, 1993. Wood-based panels: Determination of Modulus of Elasticity in Bending and of Bending Strength. The British Standards Institution, London, p. 67–90.

9. Aigbodion, V.S., Hassan, S.B. and Agunsoye, J.O., 2011. Effect of bagasse ash reinforcement on dry sliding wear behaviour of polymer matrix composites. *Materials and Design*, 33: 322–327.

10. Wasylciw, W., 1999. The utilization of industrial hemp stalks in composite panels. *Proceedings of the Meeting of the Eastern Canadian Section of the Forest Products Society*. May 19–20, Winnipeg, Manitoba

11. Ntalos, G.A. and Grigoriou, A.H., 2002. Characterization and utilisation of Wine prunings as a wood substitute for particleboard production. *Industrial Crops and Products*, 16(1): 59–68.

12. EN 312–2, 1996. Particleboards–Specifications–Part 2: Requirements for general purpose boards for use in dry conditions. *European Standardization Committee*, Brussell.

13. EN 312–3, 1996. Particleboards–Specifications–Part 3: Requirements for boards for interior fitments (including furniture) for use in dry conditions. *European Standardization Committee*, Brussell.

14. EN 312–4, 1996. Particleboards–Specifications–Part 4: Requirements for load–bearing boards for use in dry conditions. *European Standardization Committee*, Brussell.

15. EN 312–6, 1996. Particleboards–Specifications–Part 6: Requirements for heavy–duty load–bearing boards for use in dry conditions. *European Standardization Committee*, Brussell.

16. Razali, A.K. and Kuo, H.S., 1991. Properties of particleboards manufactured from fast growing plantation species. In: *Proceedings of a Regional Symposium on Recent Developments in Tree Plantations of Humid/Subhumid Tropics of Asia*, (Eds.) S.A. Abod *et al.*, 5–9 June 1989, Serdang. Universiti Putra Malaysia, Serdang, pp. 685–691 .

17. Anonymous www.finsa.com, 2010–03–10.

18. Anonymous www.spanogroup.be, 2010–03–01.

19. Anonymous www.linex–pg.nl, 2010–03–04.

Chapter 13

Sol-Gel Synthesis of Nanocrystalline Anatase Titanium Dioxide (TiO_2) at Different pH Values and Temperatures

Amir Habib[1] *and Shahid Iqbal*[2]

[1]*School of Chemical and Materials Engineering, National University of Sciences and Technology, H-12, Islamabad 44000 Pakistan, 0092 300 9756628*

ABSTRACT

The sol-gel wet chemistry route was used for the synthesis of TiO_2 nanoparticles at different pH ranges, by using simple reaction conditions with simple experimental setup. TiO_2 nanoparticles (~4-13nm) were synthesized by the reaction of titanium tetra chloride and water using HCl and NaOH as catalysts at room temperature. The precipitates obtained were washed, dried, and calcined for characterization using X-Ray Diffraction, Scanning Electron Microscopy, Transmission Electron Microscopy, Energy Dispersive X-Rays and X-Ray Fluorescence for their phase identification, structure, size and morphology, and elemental composition. Change in synthesis parameters like temperature and pH affected the particle size, morphology and phase composition. The samples synthesized in acidic conditions became crystalline at very low temperatures. The particle size remained very small (3-6 nm) in the acidic conditions. No crystallinity was observed under alkaline conditions

---

E-mail: 1: amirhabib@scme.nust.edu.pk, 2: sij_jadoon@hotmail.com

pH 8-12. The crystallize size increased with increase in pH and increase in calcination temperature. The suitable, pH range was found to be 2-6 at a calcination temperature of 200°C to get spherical anatase TiO_2 water dispersible nanoparticles with uniform size distribution.

Keywords: *Sol gel synthesis, TiO_2 Anatase, Nanoparticles, X-Ray Diffraction, Scanning electron microscopy, Transmission electron microscopy.*

Introduction

Titanium dioxide, also known as Titania is a low cost, ninth most common mineral on earth having high refractive index ($\eta \sim 2.3$–2.9) (Inoue, Muto *et al.,* 2004), photocatalytic activity, strong oxidizing power, nontoxicity, long-term photostability (Calvo, Colodrero *et al.,* 2008) and excellent electron transfer properties (Klimov 2004). Titanium dioxide occurs in three most common and most abundant crystallographic forms of brookite (orthorhombic), rutile (tetragonal structure) and anatase (body centered tetragonal) (Madhusudan Reddy, Manorama *et al.,* 2003; Thevenot, Cho *et al.,* 2008). The most commonly studied TiO_2 photocatalyst Degussa P25 standard material is a three-phase mixture of the amorphous, anatase and rutile forms of TiO_2, anatase constitutes the greatest volume fraction of approximately 80 per cent of this mixture (Kutal, Serpone *et al.,* 1993).

Nanocrystalline TiO_2 has potential applications in many areas such as in electronics, catalysis, photocatalysis (Garzella, Comini *et al.,* 2000), fabrication, fuel cells, solar cells, in air and water purification systems, in photoelectrical cells, or as gas sensors (Rodríguez and Garcia 2007). TiO_2 has also been used as a disinfectants, in sterilization and in dental bleaching products (Eremenko, Bezuglaya *et al.,* 2001; Choi, Mamak *et al.,* 2004). All these applications depend upon the ability of nano-crystalline TiO_2 to perform photocatalysis, which in turn depends on the particle morphology, average size, and phase composition.

Nanosized anatase TiO_2 is most attractive for these applications because of its large effective surface area which enhances the catalytic activities especially in photocatalysis (Jiang and Gao 2003; Bakardjieva, Subrt *et al.,* 2005). The anatase is most difficult to synthesize (Madhusudan Reddy, Manorama *et al.,* 2003). The purposes of the work is to develop a synthetic route that permit to get specific phases, controlled surface area, specific morphological features and crystals dimensions at low temperature and at specific pH value. The TiO_2 nanoparticles synthesized via most of the traditional routes are amorphous. Consequently, additional calcinations are generally required to induce crystallization, which frequently leads to the particle growth and agglomeration. Earlier researchers mostly focused to study the effects of temperature, solvent/precursor ratios and types (Park, Kim *et al.,* 1997), and effects of concentrations of reagents, gaseous atmosphere and varying calcinations temperatures and in a few cases aging period and drying conditions. Many researchers focused to study and enhance the photocatalytic activity of TiO_2. Some researchers synthesized the TiO_2 nanoparticles at a particular pH (Zhang, Gao *et al.,* 1999; Madhusudan Reddy, Manorama *et al.,* 2003; Bakardjieva, Subrt *et al.,* 2005). But these works don't discuss the effect of pH and calcination temperature (Deng, Cheuk *et al.,* 2007) on particle size and morphology. The previous works involved prolonged mixing of precursors, long stirring time, solvent removal at higher temperatures (Porkodi and Arokiamary 2007), presence of mixed phase (Karami 2010), and uncontrolled precipitation. Some used surfactants as morphology controllers, but surfactants removal from the final product become a problem and it causes impurities in the final product (Liao and Liao 2007). The preparation of TiO_2 nanoparticles in solution phase would be one of the best synthetic routes. In the present study we made an attempt to synthesize nanocrystals of pure 100 per cent anatase TiO_2 by sol-gel method at low temperature while others reported mixed phases. The present work deals with the influence of the most important parameter that is pH, on the physio-chemical properties of nano TiO_2.

Experimental

Reagents and Chemicals

The raw materials were Titanium tetrachloride (98 per cent $TiCl_4$), Sodium Hydroxide (NaOH), Hydrochloric acid (37 per cent HCl) and deionized water. All chemicals used in experimental work were pure and analytical grade reagents of noted firms such as Merck, Fluka and Aldrich and were used without any further purification. Titanium tetrachloride was used as a precursor for Titanium. Careful handling in dry atmospheres was required to avoid rapid hydrolysis and uncontrolled precipitation because most metal halides are highly reactive towards water. During experiment, dry conditions were applied to avoid reaction of $TiCl_4$ with air and humidity. The simple laboratory glassware's were used such as volumetric and titration flasks, condenser, three-neck round bottom flask, beakers, thermometer and pH meter etc.

Procedure

The 10ml of HCl was added to 250ml of deionized water to get 0.5 M acidic solution of pH 0.5. Then 0.5M stock solution for Ti^{+4} was prepared by adding 14.06ml of $TiCl_4$ in the 235ml of acidic solution (HCl + deionized H_2O) in 3-neck round bottom flask. The solution was made acidic to avoid the explosive generation of orthotitanic acid $Ti(OH)_4$ formed when $TiCl_4$ was added to this solution. The pH of the stock solution was reduced due to liberation of HCl gas produced as a result of reaction of $TiCl_4$ with water. Condenser was fitted in middle neck to condense any vapors produced during the course of reaction, also to avoid the bursting of 3-neck flask, due to liberation of HCl gas. The $TiCl_4$ was added drop by drop from the dropping funnel fitted in neck of 3-neck round bottom flask and in the third neck of round bottom flask quick fit thermometer was inserted to trace the temperature. The 3-neck round bottom flask was placed in a water bath containing ice to avoid the heat generated during the hydrolysis reaction. The assembly was placed in Fume Hood. The solution was stirred to avoid formation of any precipitate. While dropping $TiCl_4$ drop by drop the color of the solution turned yellowish, after dropping all the $TiCl_4$, the solution became colorless after sometime. The stock solution was prepared at room temperature.

Then 30ml from the stock solution was taken in a titration flask and was neutralized with 0.5M solution of Sodium Hydroxide till the pH reached to the desired value *i.e.*, from pH=2 to pH=13. While neutralizing the pH was monitored by Martini pH meter and the solution was continuously stirred vigorously with magnet stirrer. TiO_2 gel formed was filtered using Buckner assembly fitted with vacuum pump at room temperature. The Whatman filter paper (no. 42) was used. After removal of solvent, gel was washed with distilled water several times to remove any impurities that were mostly sodium and chloride ions. The samples synthesized at specific pH values were taken in pettry dishes and kept in oven at 100°C, 150°C, 200°C and 250°C

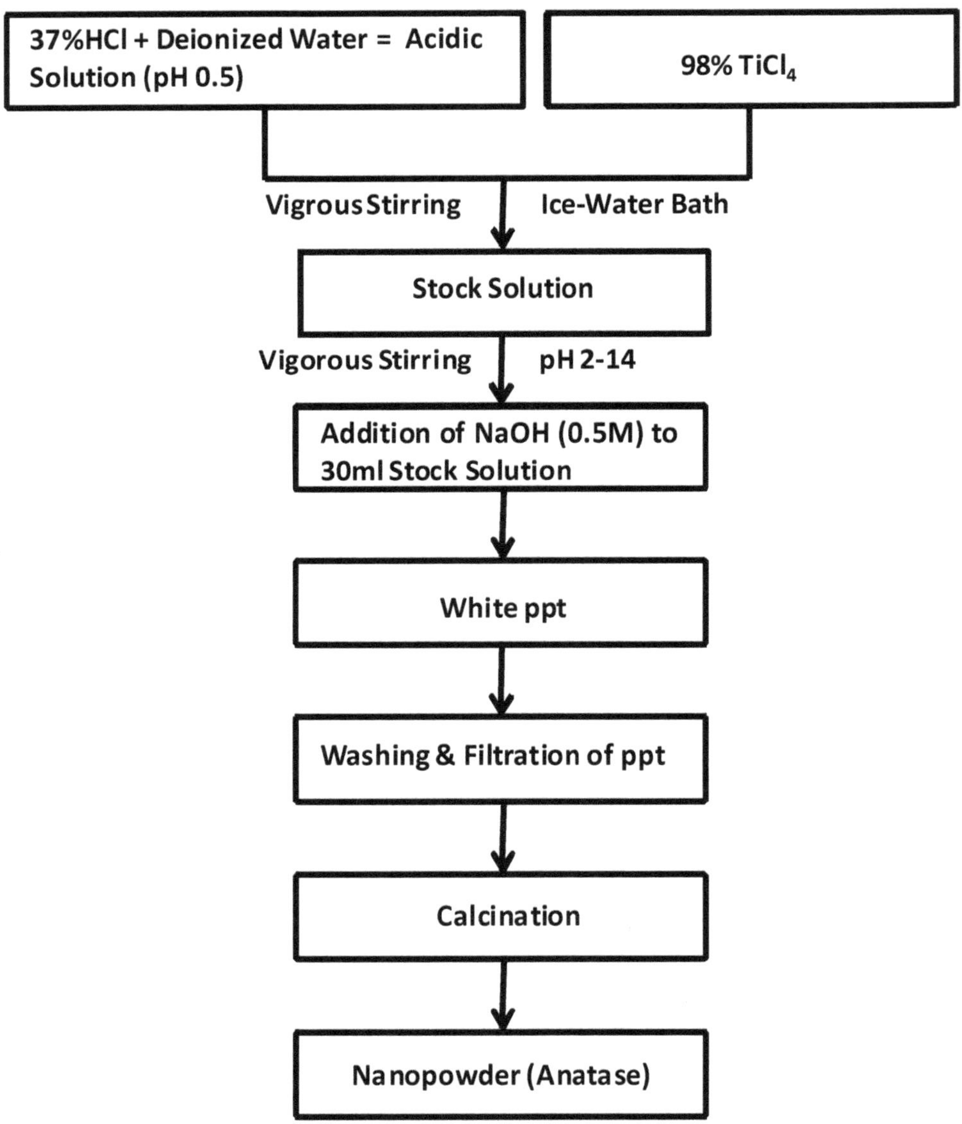

Figure 13.1: Schematic Procedure for Synthesis of TiO_2 Powders via Sol-Gel Route Used in this Study

for five hours. After five hours calcination, the samples were taken out of oven and kept at room temperature. Schematic procedure for TiO_2 synthesis is provided in Figure 13.1. Then samples were grounded with mortar and pestle. The powder samples obtained were then saved in sample vials for characterization. The sol-gel-derived powders were then characterized by X-Ray Diffraction (XRD), Scanning Electron Microscopy (SEM), X-Ray Fluorescence (XRF) and Transmission Electron microscope (TEM) techniques. The synthesized samples are given in Table 13.1.

Table 13.1: TiO_2 Samples Obtained at Different pH Values and Calcinations Temperatures for 5 Hours Each. The blanks indicate the samples were not synthesised due to expected amorphous nature of the product.

Sample at (100°C)	Sample at (150°C)	Sample at (200°C)	Sample at (250°C)
A-pH02	B-pH02	C-pH02	--
A- pH03	B- pH03	C- pH03	D- pH03
A-pH04	B-pH04	C- pH04	D- pH04
A-pH05	--	C- pH05	D- pH05
A-pH06	--	C- pH06	--
A-pH07	--	--	--
A-pH08	--	--	--
A-pH09	--	--	--
A-pH10	B-pH10	C- pH10	C- pH10
A-pH11	--	--	D- pH11
A-pH12	B-pH12	C- pH12	D- pH12
A-pH13	B-pH13	C- pH13	D- pH13
A-pH14	B-pH14	--	--

X-Ray Diffraction (XRD)

XRD was done using X-ray diffractometer [Siemen's/D5000 diffractometer with software package "WinXpow"] using Cu $K\alpha$= 1.5406 Å radiation with maximum output 60KV and 60 mA. Absolute scan with step mode was used, the starting 2θ angle was 20° and end angle was 80° with step size = 0.025° and time per step was 2s with incident beam width 15 mm at 40 KV and 45 mA.

Scanning Electron Microscope (SEM) and Energy Dispersive X-rays (EDX)

High Resolution Scanning Electron Microscopy (64904-Analytic Scanning Electron Microscope JEOL Japan) was used to observe the particle size and morphology of sample powders. The chemical composition of obtained samples was as well analysed with Energy Dispersive X-Ray (EDX) Analysis. The obtained sample powders were subjected to ultra-sonication for 10 min in deionized water to disperse the particles. The dispersed particle were dropped on Al tape or in some cases on glass slide and dried. A coating of gold layer of thickness about 250Å, using Ion Sputtering Device (JFC-1500, JEOL Japan Quick Auto Coater) was done to make samples conductive.

Transmission Electron Microscope (TEM)

High resolution transmission electron microscope (HRTEM Model H-600 Hitachi using the accelerating voltage of 300 kV) was used.

X-Ray Fluorescence (XRF)

The XRF analysis was performed using JSX-3201M (JEOL Japan).

Results

X-Ray Diffraction

The phase composition and crystallite size of as-prepared TiO_2 samples were determined by X-ray diffraction. Figure 13.2 shows the XRD patterns for various samples calcined at 100°C obtained at pH range 2- 14. It is obvious from the peaks obtained in the XRD of samples that obtained powders are not fully crystallized, and suppressed peaks of Anatase TiO_2 with (101) peak is appearing in all the samples in the pH range 02-09 (a), whereas in case of basic conditions the sample are fully amorphous as no peaks are observable for pH range 10-13 (b).

Figure 13.3 (a) shows XRD patterns of samples obtained at various pH calcined at 150°C, it is observed that the samples synthesized in extremely acidic range *i.e.*, pH 2–4, are crystalline TiO_2 Anatase phase with (101) peak, while in extremely basic conditions *i.e.*, pH 10 - 14 no crystallinity is observed and totally amorphous phase is obtained. Similarly in Figure 13.3 (b) samples calcined at 200°C in acidic range show the presence of crystalline TiO_2 Anatase phase with (101) peak and basic conditions yield amorphous TiO_2. Figure 13.4 shows the same observation for samples obtained at various pH and calcined at 250°C.

The crystallite size measurement is performed using Scherrer's equation for (101) peak for the crystalline TiO_2 samples. The crystallite size of obtained TiO_2 powders for different sample is in the range of 4-13 nm (Figure 13.5). The specific surface area of the samples was calculated using the formula: $S= 6 \times 10^3/\rho L$, Where, S is the specific surface area ($m^2\ g^{-1}$), L is the average crystallite size, and ñ is the density of the titania (3.9 g cm^{-3}). It remained in the range 3.84 - $1.18 \times 10^5\ m^2g^{-1}$.

Scanning Electron Microscopy and Energy Dispersive X-Rays

SEM images of samples obtained after calcination at 100°C indicate that the particle size was in nano domain when the pH was in the acidic range, in basic range no definite particle morphology was observable, and instead a net of condensed gel was obvious. Agglomerated particles could be seen in case of samples obtained after heating the obtained gel at 150°C in the acidic and basic range of pH (Figure 13.6).

Samples obtained after calcination at 200°C and 250°C at acidic and in moderately acidic pH range were characterized by SEM. It can be seen from Figure 13.7 that well dispersed nanoparticle (~10-20 nm) with well-formed spherical morphology were obtained at 200°C in the pH range 04-06. The size of the particles is

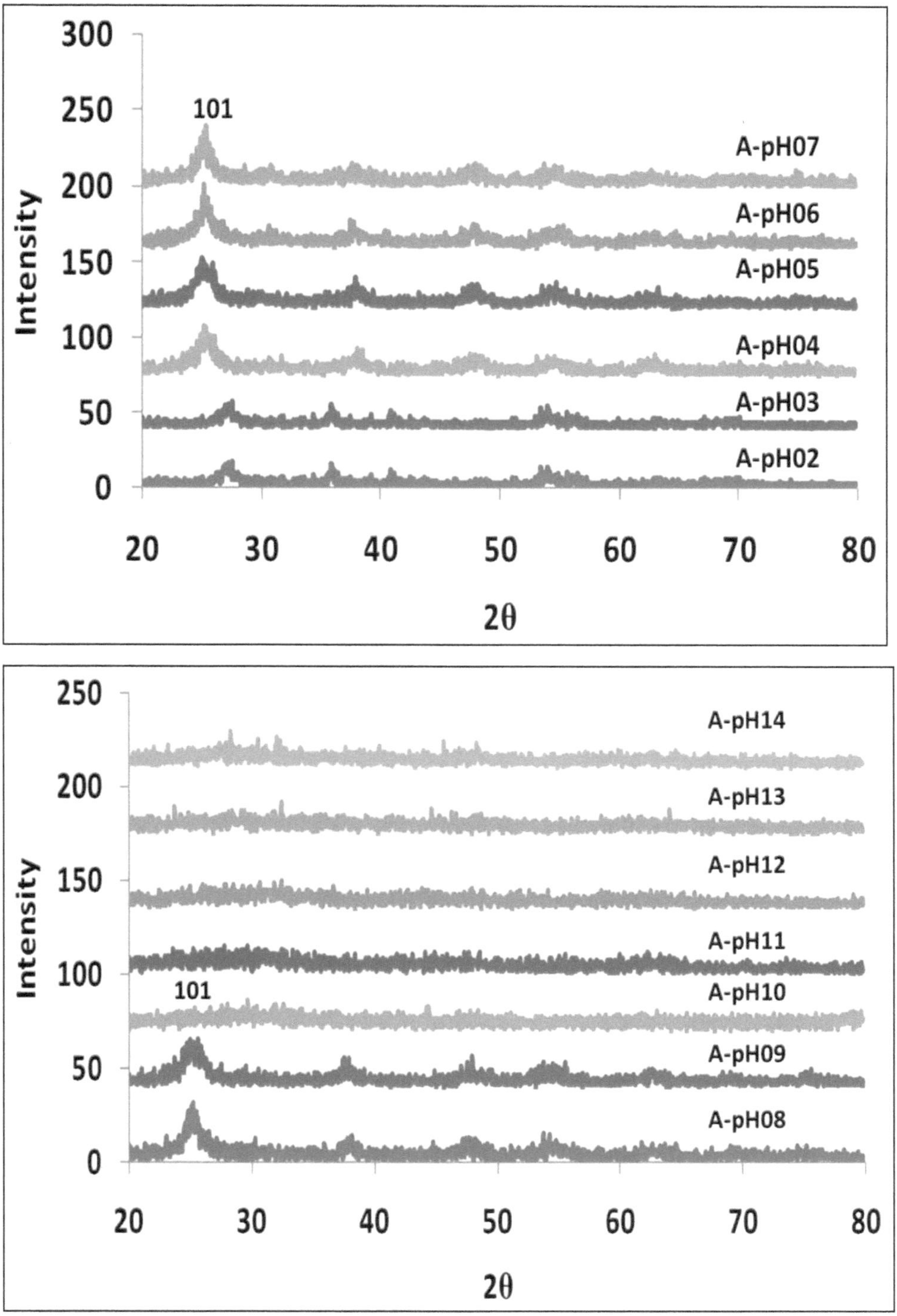

Figure 13.2: XRD Spectrums of Powder Samples Obtained at Various pH and Calcined at 100°C

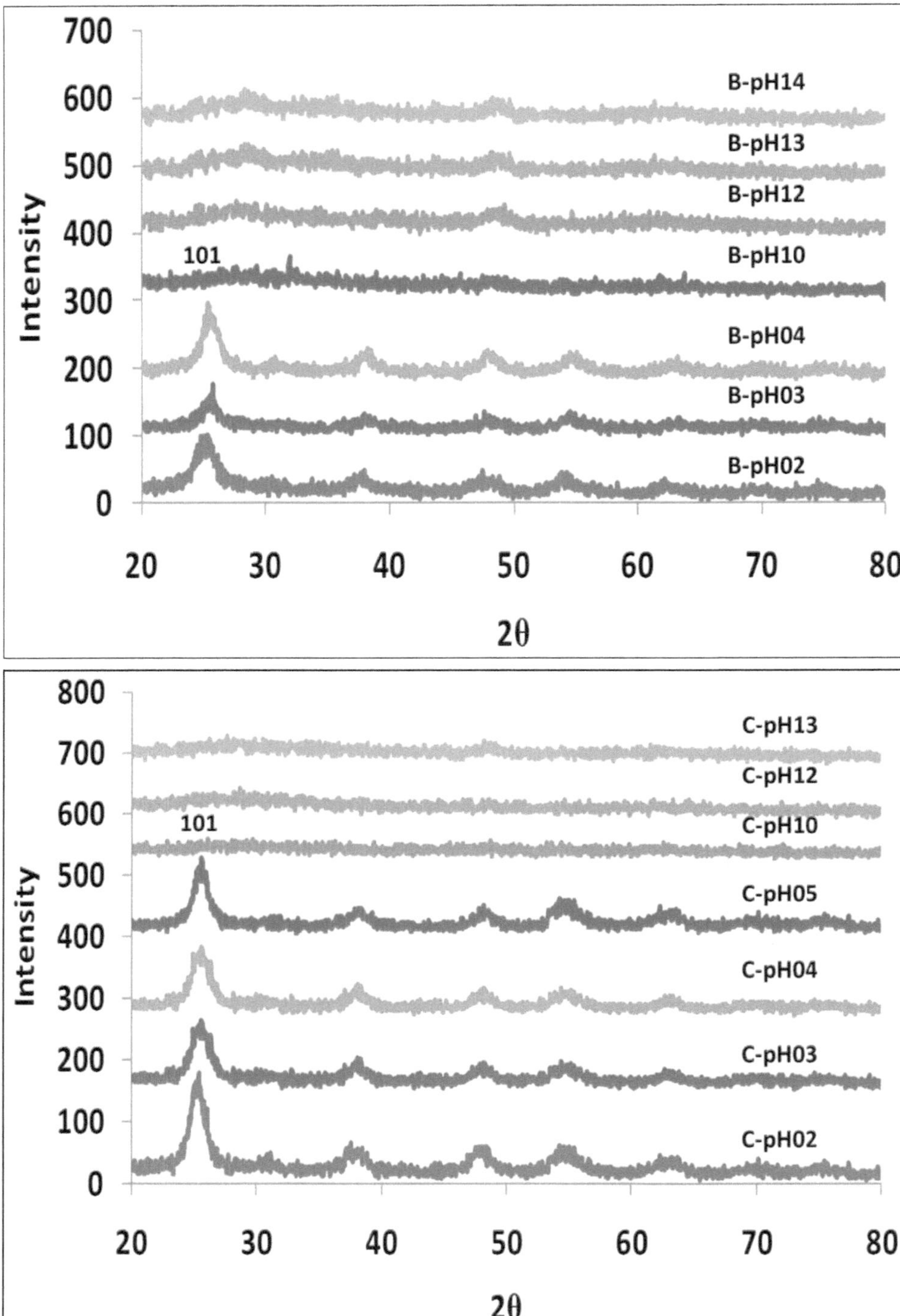

Figure 13.3: XRD Spectrums of Powder Samples Obtained at Various pH and Calcined at (a) 150°C and (b) 200°C

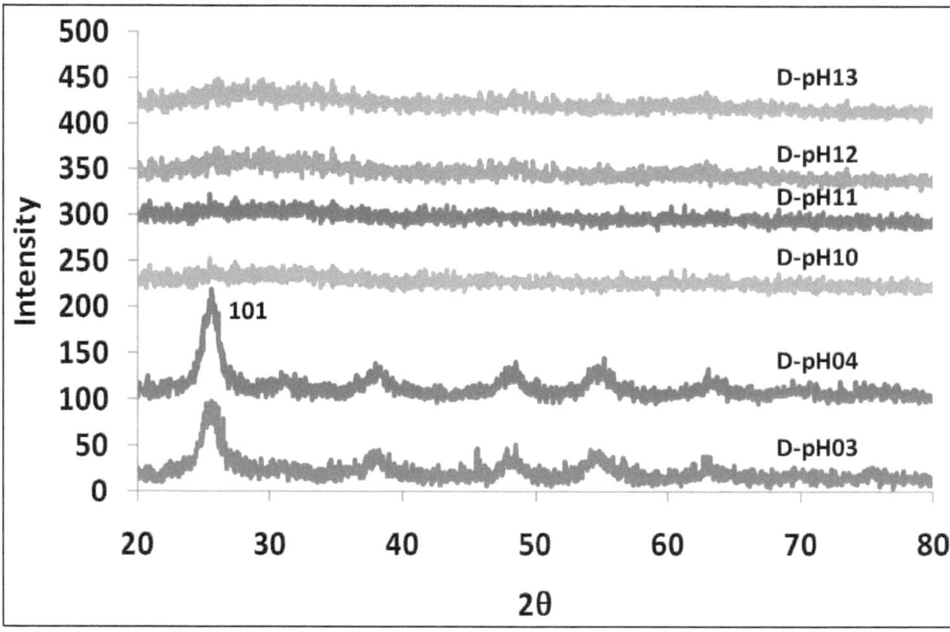

Figure 13.4: XRD Spectrums of Powder Samples Obtained at Various pH and Calcined at (a) 250°C

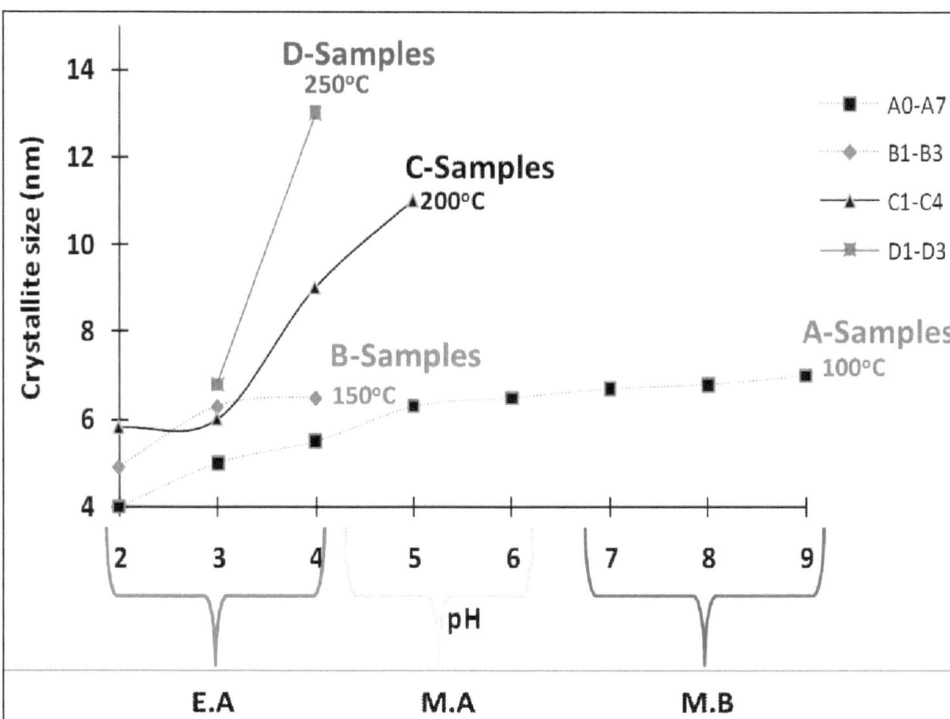

Figure 13.5: Variation of Crystallite Size with pH at different Temperatures

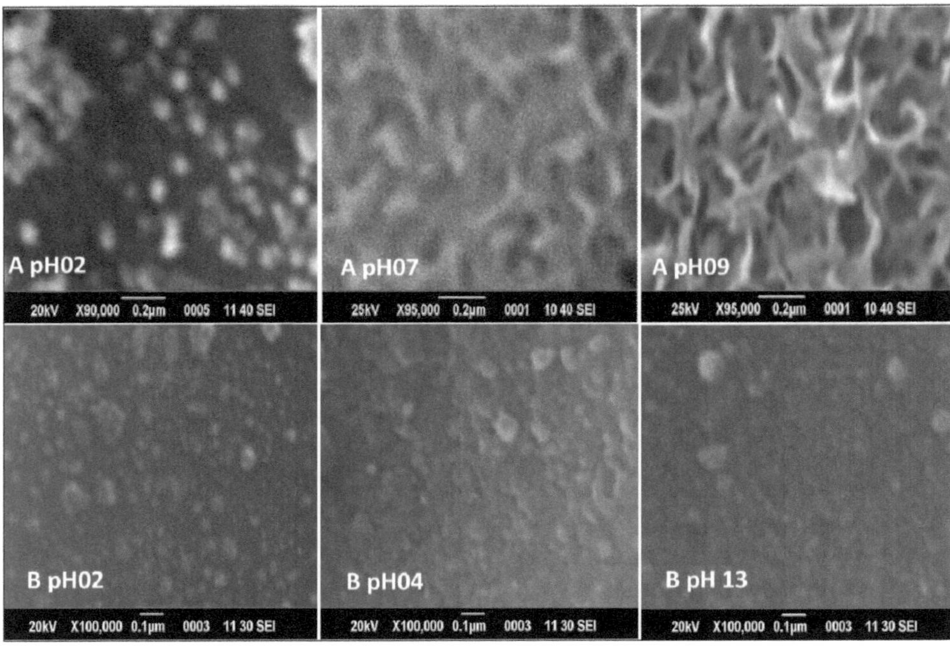

Figure 13.6: SEM Image of Samples Obtained at different pH Values and Calcinations Temperature of 100°C (A set of samples) and 150°C (B set of samples) for 5 hours

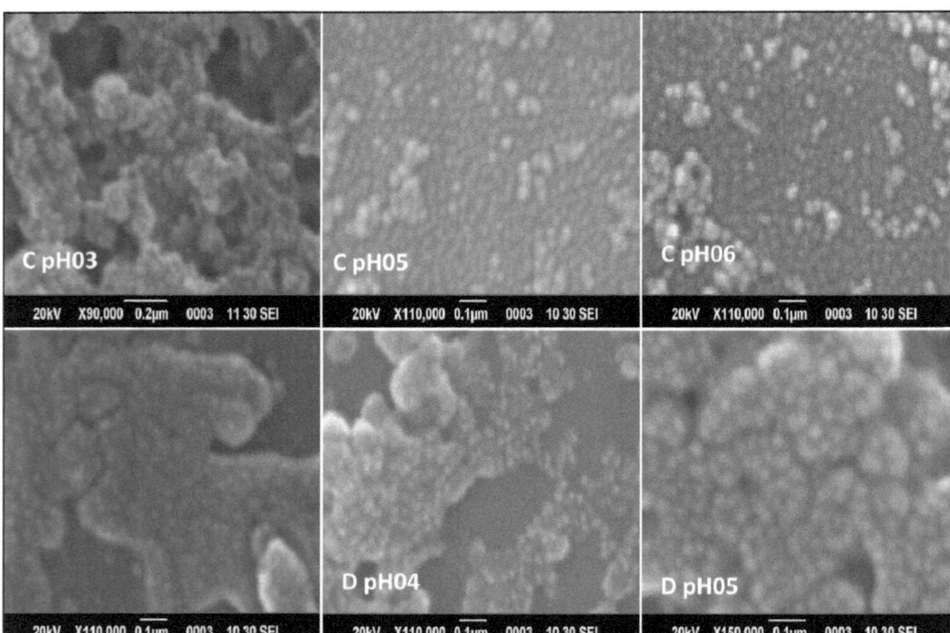

Figure 13.7: SEM Image of Samples Obtained at different pH Values and a Calcinations Temperature of 200°C (C set of samples) and 250°C (D set of samples) for 5 hours each

nearly homogenous and particles are well packed on the surface of glass substrate like beads. In case of samples obtained at even higher calcination temperature that is 250°C, the particles are formed yet they are agglomerated within the clusters.

The EDX results (Figure 13.8, Table 13.2) indicate that there are no impurities, except chlorides and sodium in some cases, in trace amounts. The results show that Titanium and Oxygen are in appropriate ratios. The signals are detected for titanium and oxygen, together with the signal coming from the gold coating and substrate due to the small thickness of the film.

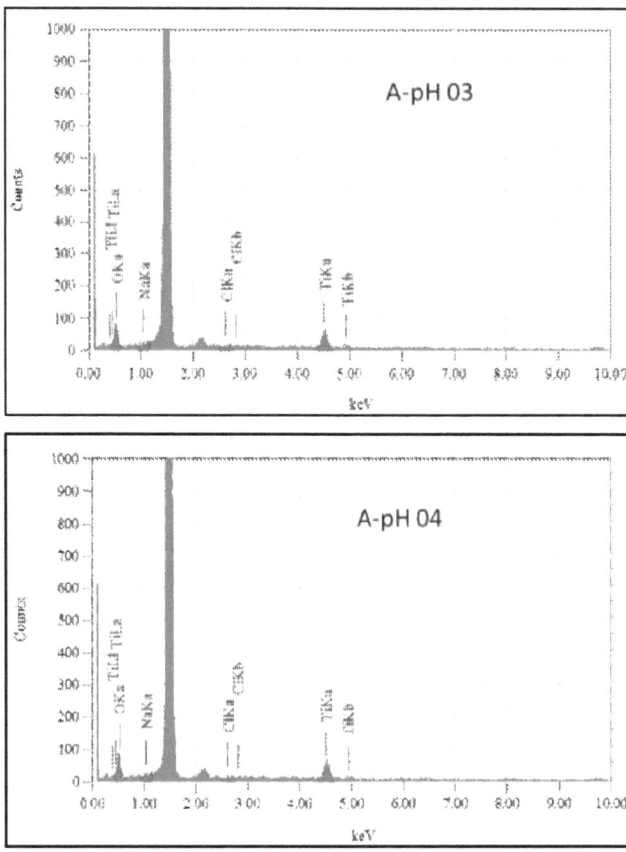

Figure 13.8: Qualitatively EDX Analysis for Sample A at Two pH Values 03 and 04

X-Ray Fluorescence

Figure 13.9 shows XRF spectrum of sample A at two pH values 02 and 04. These results are just representative for other samples. In all the samples it is observed that there are no impurities except in few samples chlorides were found in trace amounts.

Transmission Electron Microscopy

The Figure 10 shows TEM images of a sample obtained in acidic range pH 03 calcined at 100°C for five hours. The single crystal nanoparticles in the range of 5 to

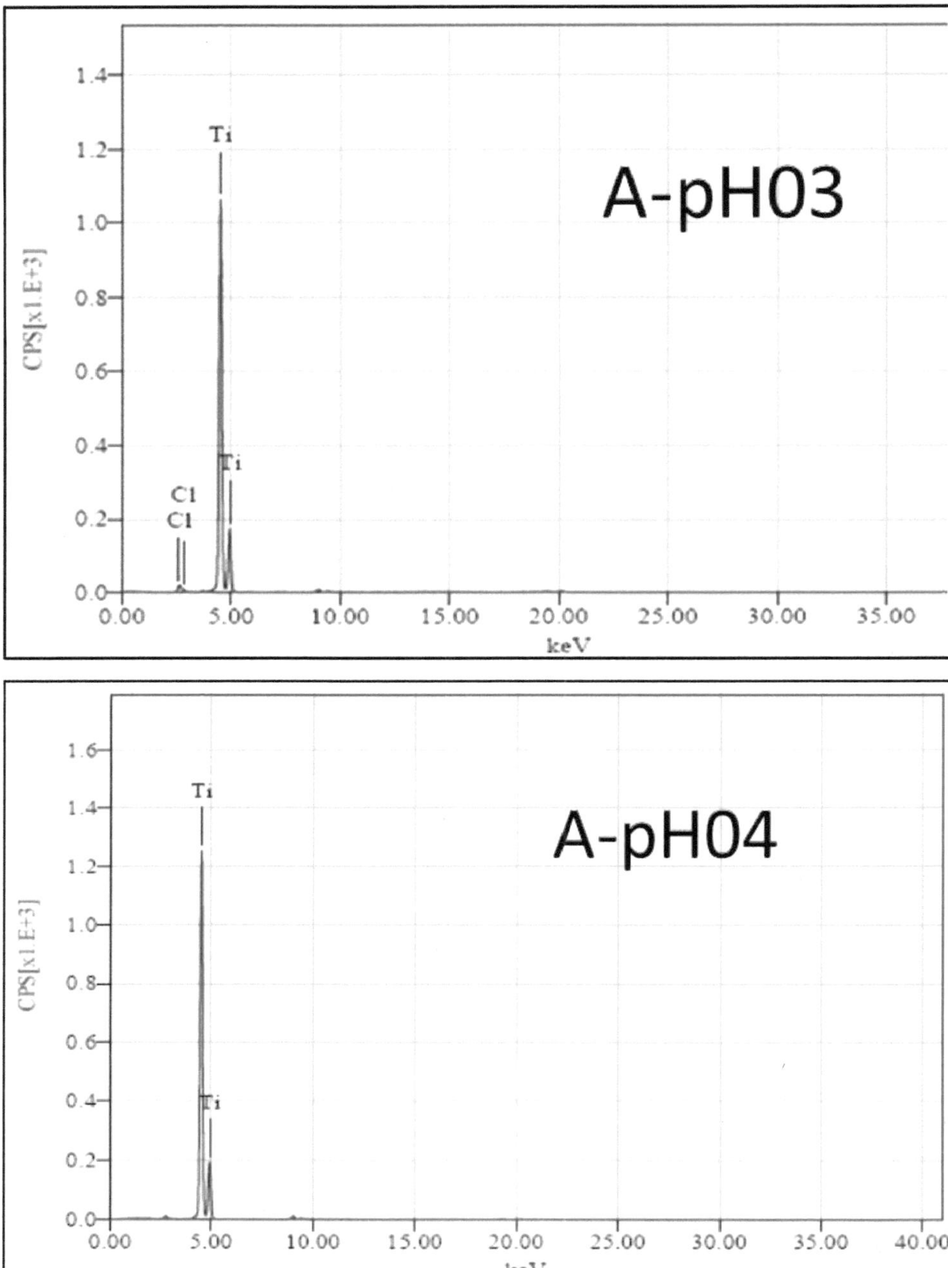

Figure 13.9: X-Ray Fluorescence Spectrum of Sample A at Two pH Values 03 and 04

10 nm are appearing with uniform size distribution. The nanoparticles of TiO_2 have elongated morphology. Figure 13.10 (c, d, f and g) show crystal lattice planes of nano-TiO_2 in anatase phase.

Table 13.2: Energy Dispersive X-Ray Results for Samples Obtained at different pH and Calcinations Temperature of 100°C

Samples	EDX analysis (Atomic wt. %)						
	Ti	O	Cl	Si	Fe	Na	S
A- pH05	89.46	10.07	0.05	0.00	0.00	0.42	0.00
A- pH06	80.00	20.00	0.00	0.00	0.00	0.00	0.00
A- pH07	81.48	18.52	0.00	0.00	0.00	0.00	0.00
A- pH08	80.60	19.40	0.00	0.00	0.00	0.00	0.00
A-pH09	81.00	19.00	0.00	0.00	0.00	0.00	0.00

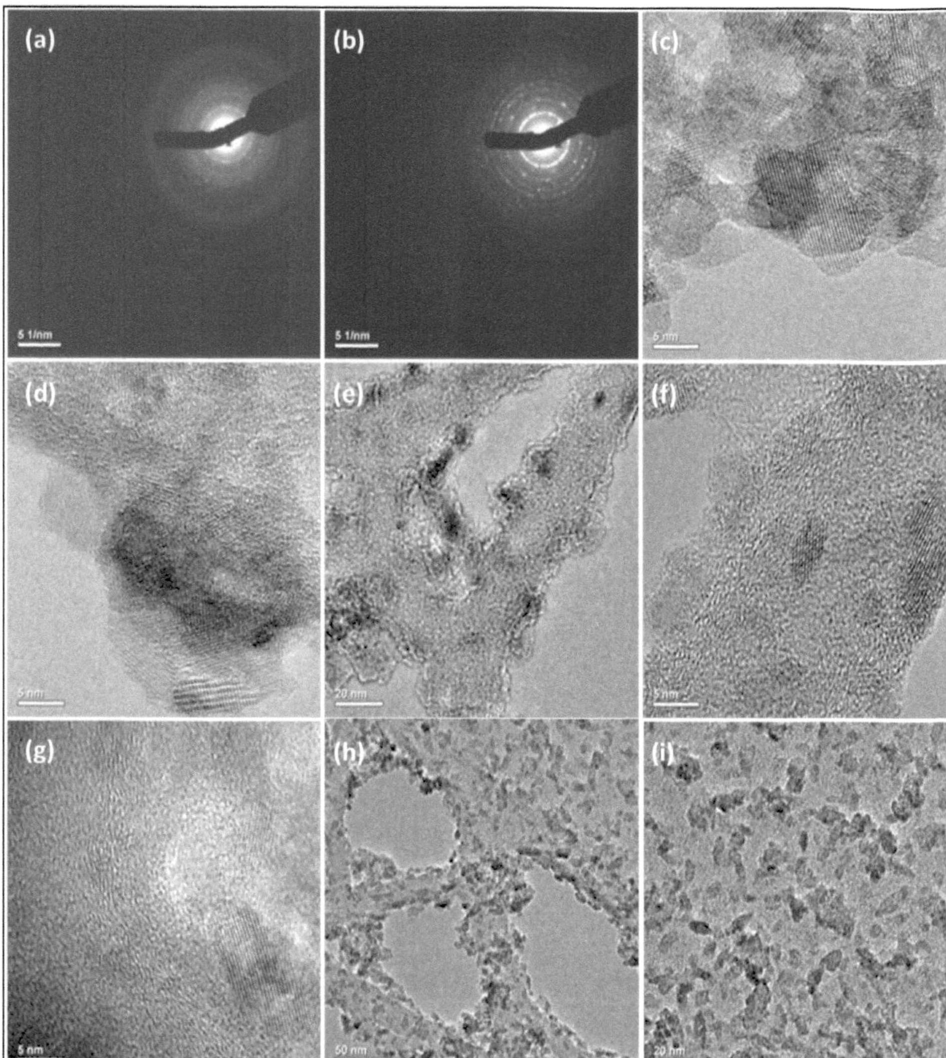

Figure 13.10: TEM Analysis of a Sample Obtained at pH 03 and Calcined at 100°C for 5 Hrs

Figure 13.11 shows the TEM images for a sample obtained in the basic range pH13 calcined at 100°C for five hours; here agglomerated amorphous particles are evident. The particle morphology is as well not obvious.

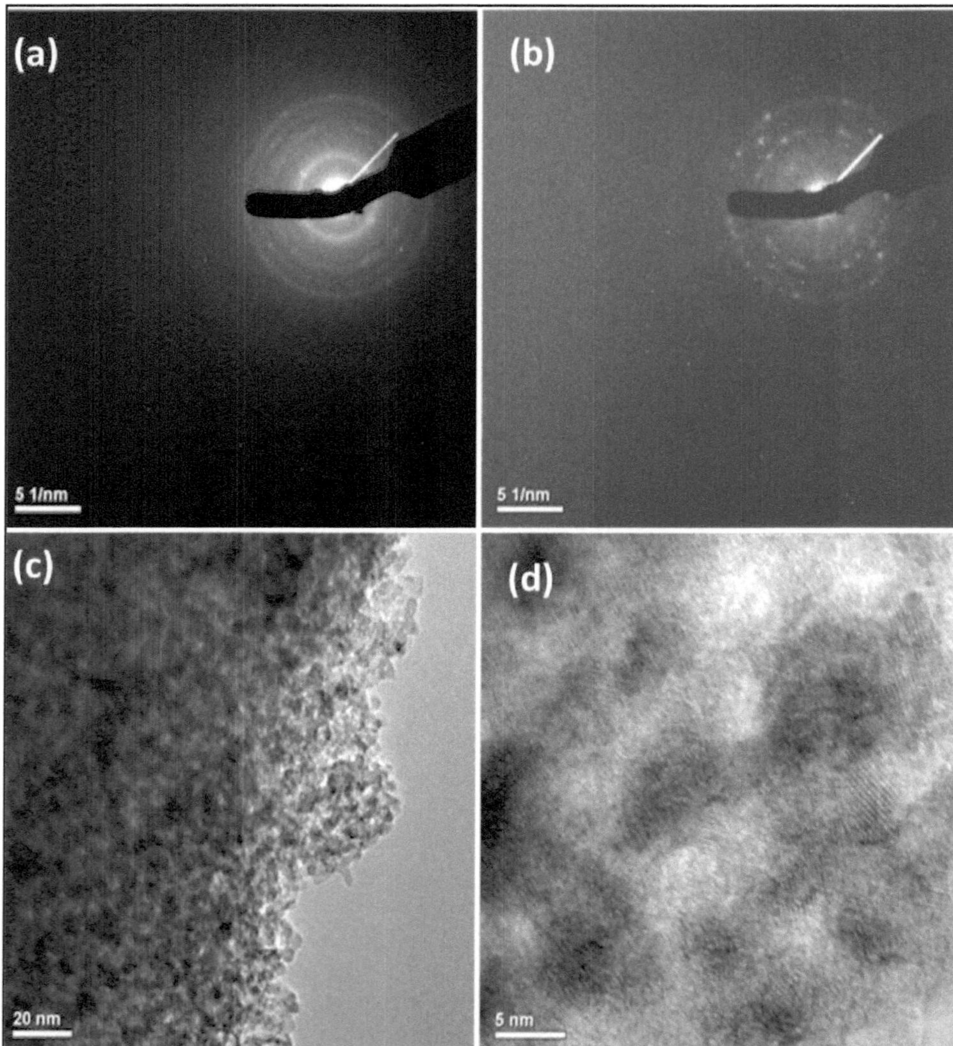

Figure 13.11: TEM Analysis of a Sample Obtained at pH 13 and Calcined at 100°C for 5 Hrs

Discussion

Effect of pH on the Formation of Anatase

Solgel process yields TiO_2 nanoparticles by hydrolysis of $TiCl_4$ and condensation of the metal hydroxide complexes formed during hydrolysis. The solgel method provides control over the reaction parameters. The rates of hydrolysis and condensation play a key role in imparting crystallinity and specific morphology to obtained product. The pH value of the sol-gel is a decisive factor for controlling the final particle size, shape, phase and agglomeration (Bakardjieva, Subrt *et al.*, 2005). The rate of hydrolysis and condensations depend upon the pH of solution and temperature. The slow and more controlled hydrolysis leads to smaller particle sizes (Rodríguez and Garcia 2007).

While preparing the stock solution (~ pH 0.5), a great care has been take that uncontrolled hydrolysis does not occur. Addition of $TiCl_4$ into water is an exothermic reaction; therefore the stock solution is prepared in a flask within a container filled with chilled water and ice. This helps to avoid the formation of Ti $(OH)_4$ polymeric network during initial stages of reactions due to heat release. During the process of crystal growth, acid works like a chemical catalyst to cause a change in the crystallization mechanism. The acidic solution hinder the rapid condensation reaction that could lead to the formation of a dense inorganic network, uncontrolled size and phase composition yielding poorly structured materials. Using these dry conditions and acidic solution, the hydrolysis and condensation speed of the metal ion precursor becomes quite low. Ti $(OH)_4$ links together to form a three dimensional lattice of linked TiO_2 forming the gel in a process called condensation. The base provides the OH^{-1} ions to the Ti^{+4} cations to form Ti $(OH)_4$ network.

XRD spectrums indicate that all the samples synthesized in the "extremely basic pH range" show no crystallinity while the samples synthesized in "extremely acidic, moderate acidic and moderate basic pH ranges" are crystalline with Anastse phase stabilization. In this study, the anatase is obtained at pH 2-9. Low pH value (high H^+ concentration) favors the formation of anatase. According to the literature pure anatase phase can be synthesized at pH 3(Li, Karger Kocsis *et al.*, 2009).

Factors Effecting the Stabilization of Anatase TiO_2

The anatase formation and stabilization may be attributed to following factors as well;

1. The brookite phase is not formed because, it is stable only at very low temperatures (below 100°C) (Yin, Li *et al.*, 2002; Guo, Gao *et al.*, 2008).
2. Rutile formation needs the calcinations temperature above 400°C. The rutile phase is obtained at higher reaction temperatures than the anatase phase.
3. Use of TiCl4 as well helps stabilize the anatase phase as the chloride ions accelerate the nucleation of anatase.
4. Size of the nanoparticle of TiO_2 obtained in this work in below 15nm. Surface free energy depends on the size. Anatase (101) is obtained, because (101) surface of anatase is the most thermodynamically stable, having lowest energy, when particle size is below 15nm. But when size is between 15 and 35 nm the most stable phase is brookite, whereas rutile is the most stable for crystal bigger than 35 nm (Pokrant and Irsen 2008; Alammar, Birkner *et al.*, 2010).

Effect of pH on the Crystallite Size of Anatase

Figure 13.5 indicates the crystallite size is increasing as pH value is increasing. The crystallite size of the samples synthesized at extremely acidic pH range (i.e pH=2-4) is smaller than those synthesized at moderate acidic pH range (*i.e.* pH=5-6) and moderate basic pH range (i.e pH=7-9). Appearance of sharp peaks for "B", "C", and "D" samples indicate that the nanoparticles are highly crystalline. In acidic conditions, crystallinity is high but in alkaline conditions the nanoparticles loses crystallinity and amorphous phase is obtained.

At pH=7-9, the samples are crystalline. It is because the OH^- ions concentration is gradually increasing after pH=7. In this pH range there are still H^+ ions that are enough to protonate the $Ti(OH)_4$ complexes. The OH^- ions are removed from the polymeric network in the form of H_2O. But at extremely basic conditions (pH=10-14), the concentration of OH^- ions is very high, while the concentration of H^+ ions is low, that hinders the formation of crystalline phase.

Effect of Calcination Temperature

Temperature plays very important role in inducing crystallinity and specific size of the TiO_2 nanoparticles. According to the XRD results for the obtained samples (Figures 13.2–13.4), the peaks are not sharp for the samples calcined at low temperature (100°C) as compared to the samples calcined at higher temperatures (150-250°C) and calcination at higher temperatures, increases the crystallite size as well (Figure 13.5). The samples calcined at 100°C are smaller in size than the samples calcined at 150°C, 200°C and 250°C. Heating induces rapid loss of solvent from the gel that causes the TiO_2 particles to grow in size.

It is observed for the 'B', 'C' and 'D' set of samples, that with rise in temperature of calcinations, crystallinity is induced rapidly, to greater extent, with elimination of OH^- ions from the $Ti(OH)_4$ polymeric network, that's why size is increasing and intensity increases for samples. In case of "D" set of samples, the temperature of calcination *i.e.*, 250°C is not inducing crystallinity in the samples synthesized in extremely basic range. This indicates that 250°C is not enough temperature to induce crystallinity in the basic range. It is observed in case of all samples that temperature of calcinations is not inducing crystallinity in the samples synthesized in the "Extremely Basic pH Range" and also pH is not playing any role here to induce crystallinity. In order to induce crystallinity in such extreme basic condiction, high temperature of calcination is required (Saadoun, Chorfi *et al.,* 2007). But higher temperature of calcinations will increase the crystallite size and at temperature greater than 450°C anatase is converted irreversibly into rutile (Wang, Xu *et al.,* 2006).

Effect of pH and Temperature on Morphology of Anatase TiO_2

The SEM and TEM results indicate that, with the increase in pH values at specific calcination temperature, the TiO_2 particles change their morphology. In extremely acidic pH range the particles are spherical, in moderate acidic pH range they become oval like, while in moderate and extremely basic pH range, they do not have any specific morphology.

The samples synthesized in extremely acidic pH range give good dispersion in water. While the samples synthesized in alkaline conditions, *e.g.*, A pH07_A pH12 indicate that the particles have not well defined morphology and appear as network. As we are moving towards alkaline conditions there is loss in crystallinity and morphology. The particles remain agglomerated in water after sonication. In extremely basic conditions the samples, show integrated network of particles, as they are interwoven. It is because of high concentration of OH^- ions, due to which the rate of protonation ceases, and rate of removal of OH^- from $Ti(OH)_4$ network is very low.

SEM and XRD results indicate that with the increase in calcination temperature to 200°C, the well crystalline TiO_2 nanoparticle with spherical bead like morphology is obtained at moderate pH range *i.e.*, 04-06. While the particles obtained at low calcination temperature and at even higher calcination temperatures particles are agglomerated with no well-defined shape.

According to literature temperature and pH are the important key parameters in imparting specific phase, size and morphology to the nanoparticles. In the case of anatase it has been reported that synthesis under basic conditions (pH=7-14) results in small cubic-like nanocrystals, hexagonal nanocrystals, or short rod-like nanocrystals. In contrast, acidic conditions (pH < 7) have been reported to result almost in spherical particles and the growth rate of anatase is dependent on pH. In each case the final nano-morphology is dependent upon the value of the pH and hence the properties of the nanocrystals will be sensitive to pH of solution (Calatayud, Mori-Sanchez *et al.*, 2001; Eremenko, Bezuglaya *et al.*, 2001).

Effect of Water on synthesis of TiO_2

Deionized water is used as solvent, for the synthesis of TiO_2 nanopartciles. This is easily removed during calcinations, even at low temperatures. While in case of organic solvents, after hydrolysis reactions the remaining organic molecules will lead to the incomplete crystallization of Titania, If the calcination temperature for crystallization is not high enough (Li and Ishigaki 2002; Barnard, Saponjic *et al.*, 2005).

Compositional Analysis

The EDX and XRF results show that Titanium and oxygen are in stoichiometric ratios. There are no impurities. In few samples, chlorides and sodium ion impurities in trace amounts are observed. The impurities are present may be due to poor washing of the precipitates during sample preparation.

Conclusion

TiO_2 nanoparticles (~4-13nm) particles are synthesized via sol-gel route. Pure anatase phase TiO_2 with crystalline nature is obtained in the pH range 2-9 at calcination temperatures 100, 150, 200 and 250°C. With the increase in pH at different calcination temperatures, it is observed that crystallite size of obtained TiO_2 nanoparticles increases, whereas, at a certain pH, the crystallite size is observed to increase with the increase in calcination temperature. At pH values (2-6), the particle size is small and morphology is spherical. Under alkaline conditions, the samples become amorphous and loses regular morphology and show net like structure. Titanium and oxygen are present in appropriate ratios in the crystalline samples. The rate with which -OH ions are added and the rate with which protonation occurs affect the formation of particular phase with specific morphology. Hence the suitable, pH range is 4-6 and calcination temperature is 200ºC, in order to get spherical anatase TiO_2 nanoparticles with uniform size distribution and easily dispersible in water.

References

Alammar, T. and Birkner, A. *et al.*, 2010. Sonochemical preparation of TiO_2 nanoparticles in the ionic liquid 1-(3-hydroxypropyl)-3-methylimidazolium-bis (trifluoromethylsulfonyl) amide. *Materials Chemistry and Physics,* 120(1): 109–113.

Bakardjieva, S. and Subrt, J., *et al.*, 2005. Photoactivity of anatase-rutile TiO_2 nanocrystalline mixtures obtained by heat treatment of homogeneously precipitated anatase. *Applied Catalysis B: Environmental,* 58(3–4): 193–202.

Barnard, A. and Saponjic, Z., *et al.*, 2005. Multi-scale modeling of titanium dioxide: Controlling shape with surface chemistry. *Reviews on Advanced Materials Science,* 10(1): 21–27.

Calatayud, M.and Mori-Sanchez, P., *et al.*, 2001. Quantum-mechanical analysis of the equation of state of anatase TiO_2. *Physical Review,* B64(18): 184113.

Calvo, M.E.and Colodrero, S., *et al.*, 2008. Photoconducting Bragg mirrors based on TiO_2 nanoparticle multilayers. *Advanced Functional Materials,* 18(18): 2708–2715.

Choi, S.Y.and Mamak, M., *et al.*, 2004. Thermally stable two dimensional hexagonal mesoporous nanocrystalline anatase, Meso nc TiO_2: Bulk and crack free thin film morphologies. *Advanced Functional Materials,* 14(4): 335–344.

Deng, H.and Cheuk, K., *et al.*, 2007. Low temperature preparation of nano TiO_2 and its application as antibacterial agents. *Transactions of Nonferrous Metals Society of China,* 17(s1): 700–703.

Eremenko, B.and Bezuglaya, T., *et al.*, 2001. Stability of aqueous dispersions of the hydrated titanium dioxide prepared by titanium tetrachloride hydrolysis. *Colloid Journal,* 63(2): 173–178.

Garzella, C. and Comini, E., *et al.*, 2000. TiO_2 thin films by a novel sol-gel processing for gas sensor applications. *Sensors and Actuators B: Chemical,* 68(1–3): 189–196.

Guo, L. and Gao, G., *et al.*, 2008. Preparation and characterization of TiO_2 nanosponge. *Materials Chemistry and Physics,* 111(2–3): 322–325.

Inoue, S., A. Muto, *et al.* (2004). "Preparation of novel titania support by applying the multi–gelation method for ultra–deep HDS of diesel oil." Applied Catalysis A: General 269(1–2): 7–12.

Jiang, H. and Gao, L., 2003. Enhancing the UV inducing hydrophilicity of TiO_2 thin film by doping Fe ions. *Materials Chemistry and Physics,* 77(3): 878–881.

Karami, A., 2010. Synthesis of TiO_2 nano powder by the sol-gel method and its use as a photocatalyst. *Journal of the Iranian Chemical Society (JICS),* 7(154): 154–160.

Klimov, V.I., 2004. *Semiconductor and Metal Nanocrystals: Synthesis and Electronic and Optical Properties*. CRC.

Kutal, C. and Serpone, N., *et al.*, 1993. *Photosensitive Metal-Organic Systems.* ACS Publications.

Li, W. and Kocsis, J. Karger, *et al.*, 2009. Dispersion of TiO_2 Particles in PET/PP/TiO_2 and PET/PP/PP g MA/TiO_2 composites prepared with different blending procedures. *Macromolecular Materials and Engineering*, 294(9): 582–589.

Li, Y.L. and Ishigaki, T., 2002. Synthesis and structural characterization of titanium oxides and composites by thermal plasma oxidation of titanium carbide. *Thin Solid Films*, 407(1–2): 79–85.

Liao, D. and Liao, B., 2007. Shape, size and photocatalytic activity control of TiO_2 nanoparticles with surfactants. *Journal of Photochemistry and Photobiology A: Chemistry*, 187(2–3): 363–369.

Madhusudan Reddy, K. and Manorama, S.V., *et al.*, 2003. Bandgap studies on anatase titanium dioxide nanoparticles* 1. *Materials Chemistry and Physics*, 78(1): 239–245.

Park, H.K. and Kim, D.K., *et al.*, 1997. Effect of solvent on titania particle formation and morphology in thermal hydrolysis of $TiCl_4$. *Journal of the American Ceramic Society*, 80(3): 743–749.

Pokrant, S. and Irsen, S., 2008. Crystal structure of TiO_2 Nano-particles. *Microscopy and Microanalysis*, 14(2): 354–355.

Porkodi, K. and Arokiamary, S., 2007. Synthesis and spectroscopic characterization of nanostructured anatase titania: A photocatalyst. *Materials Characterization*, 58(6): 495–503.

Rodríguez, J.A. and Garcia, M.F., 2007. *Synthesis, Properties and Applications of Oxide Nanomaterials*, LibreDigital.

Saadoun, M. and Chorfi, H., *et al.*, 2007. Polymer supported porous TiO_2: Application to photo catalysis. *Physica Status Solidi*, (c)4(6): 2029–2033.

Thevenot, P. and Cho, J., *et al.*, 2008. Surface chemistry influences cancer killing effect of TiO_2 nanoparticles. *Nanomedicine: Nanotechnology, Biology and Medicine*, 4(3): 226–236.

Wang, Y. and Xu, H., *et al.*, 2006. A general approach to porous crystalline TiO_2, $SrTiO_3$, and $BaTiO_3$ spheres. *The Journal of Physical Chemistry*, B110(28): 13835–13840.

Yin, S. and Li, R., *et al.*, 2002. Low temperature synthesis of nanosize rutile titania crystal in liquid media. *Materials Chemistry and Physics*, 75(1–3): 76–80.

Zhang, Q.H. and Gao, L., *et al.*, 1999. Preparation and characterization of nanosized TiO_2 powders from aqueous $TiCl_4$ solution. *Nanostructured Materials*, 11(8): 1293–1300.

Chapter 14

Production and Characterization of Dry-Pressed Porcelain

Nazar Shawgi[1] *and Mohamed Talballa*[2]

[1]*Materials and Electronics Research Institute, National Center for Research, Sudan*

[2]*Professor of Materials Engineering, Materials and Electronics Research Institute, National Center for Research, Amarat St. 15, Khartoum, Sudan*

ABSTRACT

The aim of this work was to produce ceramic materials of high electrical resistance to be utilized as insulators in electrical appliances and as internal coatings in industrial furnaces. Low manufacturing cost has been taken as a governing criterion, which may be achieved by better process control.

In the last two decades, there has been intense interest in developing advanced materials with special properties to meet new requirements of the electrical, electronic, aeronautical, chemical, nuclear and other industries. Ceramic materials have been a major focus of this interest. The main process parameters identified are shape and size distribution of grains in the starting materials, in addition to chemical composition and process parameters. Most of the ceramics manufacturing processes studied have used 'wet' starting materials, while dry-pressing received scant attention. The present work has been an effort to remedy this.

The starting materials used in the present work have been kaolin, talc, feldspar, quartz and flint blended in specific ratios. Each mixture was powdered to three particle sizes with moisture content adjusted to (0.36 per cent). Sample were shaped into cylinders by dry-pressing at different pressures (110-170) MPa, and fired for 12 hours at (1100 to 1200) °C. Each specimen was tested for electrical resistance at (12 - 17.4) kV. Volume change varied in

E-mail: 1: nasho19@yahoo.com, 2: m2003talballa@yahoo.com

the range -9.0 to + 0.00; water absorption was in the range 6.0 to 0.36 per cent ; compressive strength varied in the range 18.3 to 8.1 MPa; the microstructure showed mainly mullite and unreacted silica.

The results indicated that the dielectric strength of the specimens decreased as the applied pressure or temperature decreased. Change in dimensions decreased with increase in pressure or temperature; the same trend was observed with water absorption. The effects on compressive strength proved to be a little complex, depending on the combination of pressure and temperature instead of separate effects.

Keywords: *Ceramics, Dry-press, Advanced materials, Mullite, Refractories.*

Introduction

Automatic dry pressing of ceramics has been practiced at least as far back as 1904. The process was developed in Germany during 1920-1930, and became known in the United States around 1930, (Kingery, 1963).

Dry pressing as applied to the production of technical ceramics, comprises of the compacting of dry granulated ceramic powder in a metallic die under high pressure. The moisture content of the granulated powder may be varied between 0 per cent to 4 per cent, and rarely more. The granules are free-flowing, but relatively non plastic. They are merely crushed and densely packed under pressure, to form a coherent compact. Automatic dry pressing can only give good results, if all steps of the process are kept under careful control. Under normal conditions and supervision, it is possible to ceramics dimensions to tolerance of ±1 per cent, with a minimum of ±0.005 inch. In special cases and with very close control considerably better tolerance can be kept (Kingery, 1963; Griffin, 1995).

The raw materials must be carefully selected; they should be accepted only after they have passed customary tests of particle size distribution, chemical analysis and other aspects pertinent to quality of the particular ceramic composition (Singer and Singer, 1979). The choice of a binder is most important. Many efforts have been expended to find suitable binders. A good binder should have a combination of some properties such as high impart strength, be non-abrasive to avoid excessive die ware, decompose or oxidize readily during firing at relatively low temperatures, etc. In electronic equipments and electrical appliances components, vast numbers of ceramic insulators that have small dimensions, *i.e.* serving as spacers sockets of vacuum tubes, switches etc. all have certain requirements in common: they are dimensionally accurate, so that they can be used in automatic assembly process, and must show uniformity of dielectric and mechanical properties to insure high quality of components of which they are part of.

For all these reasons; dry-pressing techniques of ceramics powder, come too close to meet these various requirements. Hence, the effects of process parameters on the properties of dry-pressed ceramics have been the subject of the present work. The main aim of the investigation was to study the effects of major process parameters on the properties of ceramics produced by dry-pressing.

Materials and Methods

Raw Materials

Natural raw materials, such as talk, kaolin, feldspar, quartz and flint, have been used in this study without any pretreatment. Raw materials are crushed and ground to powder and the test is done by the appropriate sieves (Ryan, 1978). The Table 14.1 shows the location, the particle size and the percentage of the materials used in preparing the experimental blend used.EM3

Table 14.1: Location (in Sudan), Particle Size and Percentage of the Natural Raw Materials Used

Material	*Location in Sudan*	*Particle Size (μm)*	*Per cent*
Talc	Atparah	90	7
Kaolin	Marawy	125	45
Feldspar	Shandi	250	23
Quartz	Sabalooga	250	20
Flint	U.K.	90	5

Processing

These components were mixed and blended by agitation manually in a steel container for a half of an hour, so that proper mixing and blending are achieved. To control the moisture content of the mixed blend to less than 5 per cent before shaping(Luikov, 1936): Sample were weighed, dried at 120 °C for 10 minutes, cooled in a desiccators and then reweighed. The calculated moisture content of the samples was found to be 0.36 per cent.

The parameters studied were: chemical composition, forming pressure (110, 140 and 170 MPa) and firing temperature (1100, 1150, and 1200 °C); nine specimens were produced and tested. A metallic cylindrical die of 5 cm diameter, and open at top and bottom, lubricated by crease, was used. The "punch" which was a piston, was utilized to press the powder firmly filling the die. A charge of the correct composition was introduced into the mold cavity, and then the mold was closed by bringing the "male" (piston or punch) and "female (mold cavity)" halves together. The dry powder is compacted by applying the specified pressure for each specimen as given by Table 14.2. A hydraulic press machine (Model: Hurson), of maximum load 160 tons, and electrically operated, was used.

Specimens are coded according to a coding system which indicates changes in the parameters, and is illustrated below.

S_{TP}

Where (S) stands for sample,(P) stands for forming pressure (MPa), and (T) stands for firing temperature (°C).

Table 14.2: Specimens' Code and Values of Parameters

P_1	P_2	P_3	T_1	T_2	T_3
170	140	110	1200	1150	1100

Firing

According to specimens' code and Table 14.2, the specimens were divided into three thermal groups (T_1, T_2 and T_3). Each specimen was placed on platinum rods in order to avoid reaction between the specimens and the internal surface of the furnace chamber. The platinum rods were placed on a silica refractory plate (12 x 12 cm) to protect the silica layer of the furnace chamber from damage resulting from reaction of melted specimens if it occurs. The firing time for each specimen was 12 hours. Electrical kiln, manufactured by Carbolite was used for the firing process, the Table (3) below gives details of its specifications.

EM6

Table 14.3: Kiln Specifications

Model	*Max Temp. (°C)*	*Max Power (kW)*	*Chamber Size (mm)*			*Approx Capacity l*	*Net Weight Kg*
			H	*W*	*D*		
RHF 17/3 M*	1700	5	150	150	150	3.4	85

M*: Manually operated.

Characterization

The mineral (raw materials) chemically analyzed by the spectrophotometer (HACH, Model DR/2000, Direct Reading), and by Atomic absorption spectrophotometer: (Perkin Elmer, Model: 2380, flame double- beam system), and also tested by the X-Ray diffraction apparatus (system: Phillips, Model: X' Pert PRO Stress XRD Analyzer, Cu-target radiation).

The following procedures were employed only in the characterization of the final products after the body preparation, shaping and firing. Five tests have been utilized:

High-Voltage Testing (Sustained Low Frequency Test)

A foster Transformer (Model: WA 182, British make) was used for high voltage testing. High voltage AC tests at 50 Hz were carried out as routine tests on low voltage (230 or 400 V) equipment. Each one of these devices was subjected to a high voltage of 1 kV+2 (working voltage). A 230 V piece of equipment may thus be subjected to about 1.5 to 2 kV. These tests were generally carried out after manufacture before installation (Lucas, 2001).

The high voltage was applied across the device under test by means of a transformer. The transformer need not have a high power rating. A resistance of about 1 ohm/volt was used in series with the transformer, so as to limit the current to about 1A, in the event of a breakdown (Liebermann, 2001).

Water Absorption Test

Dry specimens heated to 115 °C are weighed, immersed in water, brought to boil, boiled for one hour, allowed to stand in water for a further twenty-three hours period, wiped of surface water, and then weighed. The result is calculated as follows percentage (Norton, 1952):

$$\text{Water Absorption (\%)} = \frac{\text{Wt. absorbed water} \times 100}{\text{Wt. of dry specimen}} \qquad (1)$$

Compressive Strength

For this test, a hydraulic universal material testing machine (Model: WE- 100, with maximum load of 1000 KN, manufactured by Jinan Testing Machine Works-China) was used. The specimens is put under compressor piston and pressure was applied until the specimens is broken, then the pressure is recorded and released (Vincenzin, 1991). The pressure of the machine was given as a load (kN), and it can be inverted to pressure in Pascal, *i.e.*

$$\text{Compressive strength (kPa)} = \frac{\text{Load (kN)} \times \text{(sp.gr.)}}{\text{Area } (m^2)} \qquad (2)$$

Microstructure

A Nikon Microscope (Model: EPIPHOT 200, made in Japan, with a camera attachment), was used to investigate the microstructure of both fired and the unfired specimens. The fired specimens were subjected to crushing, grinding and surface polishing, and then observed under the microscope, and the microstructure revealed was photographed (Marinescu, 1999).

X-Ray Diffraction

X-Ray diffraction test was used to investigate phase changes that may occur after firing. The test was performed on the specimens before and after firing to observe any phase changes resulting from the firing process (Rahman, 2003).

Result and Discussion

Chemical Analysis

The results of the chemical analysis by the Atomic absorption technique and by the spectrophotometer are presented in Table 14.

It is concluded, from the data presented in Table 14.4, that SiO_2 ratio in the kaolin (Marawy) sample is high compared to other worlds' sources, which lie between (46 to 58 per cent) (Salih, 2005); also, Al_2O_3 ratio in the kaolin sample (18 to 33 per cent) is lower compared to other sources (29 to 39 per cent) (Singer and Singer,1979). Fe_2O_3 ratio in the kaolin sample may be high, which may increase purification cost. This impurity appears as dark brown spots in the micrographs of the final products.

Table 14.4: Chemical Analysis of the Minerals

Compound per cent	*Kaolin*	*Feldspar*	*Talc*	*Quartz*
Al_2O_3	25.52	19.3	0.6	—
SiO_2	60.33	62.0	58	97
CaO	0.45	0.34	Trace	$Al_2O_{3+}alkalis_=1.2$
Na_2O	1.26	2.00	Trace	
MgO	0.25	Trace	35	
K_2O	1.00	10.9	0.71	
Fe_2O_3	1.1	0.15	0.11	0.3
L.O.I (loss on ignition)	10.09	5.31	5.57	1.5

The percentage of K_2O, in the feldspar sample, is found to be 10.9 per cent comparing with Na_2O (2 per cent), hence this of feldspar is the type known as "potash feldspar".

The high ratio of Al_2O_3 in kaolin increase the plasticity of clay, but the vice versa with SiO_2.

Product Characterization

The following tests are applied to the final products to determine their proprieties, and to decide on suitable applications.The specimens were divided into three groups, in which the pressure was taken as varied parameter and the temperature was fixed as following:

- A. S_{11}, S_{12} and S_{13}: fired at 1200 °C and pressed at (170,140 and 110 MPa), respectively.
- B. S_{21}, S_{22} and S_{23}: fired at 1150 °C and pressed at (170,140 and 110 MPa), respectively.
- C. S_{31}, S_{32} and S_{33}: fired at 1100 °C and pressed at (170,140 and 110 MPa), respectively.

Also the specimens were divided into three groups, in which the firing temperature was taken as the varied parameter and pressure was fixed as following:

- D. S_{11}, S_{21} and S_{31}: pressed at 170 MPa and fired at (1200, 1150, 1100 °C) respectively.
- E. S_{12}, S_{22} and S_{32}: pressed at 140 MPa and fired at (1200, 1150, 1100 °C) respectively.
- F. S_{13}, S_{23} and S_{33}: pressed at 110 MPa and fired at (1200, 1150, 1100 °C) respectively.

High Voltages Break-Down Test

According to the high voltage test, the specimen was put on a conjugate Schering bridge, and the voltage was increased gradually until the specimen broke down. The results are shown on Table 14.5.

Table 14.5: High Voltage Test

Group	*A*			*B*			*C*		
Specimen	S_{11}	S_{12}	S_{13}	S_{21}	S_{22}	S_{23}	S_{31}	S_{32}	S_{33}
kV	17.4	16.2	15.45	15.78	15.78	12.4	12.0	12.84	12.0
Group	*D*			*E*			*F*		
Specimen	S_{11}	S_{21}	S_{31}	S_{12}	S_{22}	S_{32}	S_{13}	S_{23}	S_{33}
kV	17.4	15.78	12.0	15.78	15.78	12.84	15.45	12.4	12.0

In group (A), the specimen subjected to the highest pressure gave the highest dielectric (S_{11}, 17.4 KV). Thus specimens' dielectric strength decreased as the applied pressure was decreased. The same trend was observed in group (B) and (C). In group (D), the specimen fired at the highest temperature, *i.e.* (S_{11}), gave the highest dielectric strength. Thus specimens' dielectric decreased according to the decrease of the applied firing temperature. Same results were noticed in group (E) and (F).

From the previous discussion, it is seen that the highest dielectric strengths were obtained on the specimens which were pressed and fired at the highest applied pressures and firing temperatures. Furthermore, in group (A and B), specimens when tested by this test, were over flashed, put in group (C) they were punctured. So, these indicate that groups (A and B) specimens are stronger than group (C) which is weak.

Volume Change (Thickness and Diameter)

The measurements of the specimens' diameter and thickness were recorded before and after firing (Thurnauer, 1958); the change results percentage are shown in Table 14.6. It is very important to ascertain whether the ceramic bodies produced by dry pressing expand or shrink on firing under pressure. Furthermore, if changes dimension occur, do they occur parallel to the surface (diameter) or in the transverse direction (thickness).

Table 14.6: Change Percentage of the Thickness and Diameter

Group	*A*			*B*			*C*		
Specimen	S_{11}	S_{12}	S_{13}	S_{21}	S_{22}	S_{23}	S_{31}	S_{32}	S_{33}
Thickness change (– %)	0.003	0.93	2.77	1.82	5.56	6.42	2.00	9.09	7.27
Diameter change (– %)	0.002	0.40	0.60	0.60	1.00	1.00	0.80	1.40	1.60
Group	*D*			*E*			*F*		
Specimen	S_{11}	S_{21}	S_{31}	S_{12}	S_{22}	S_{32}	S_{13}	S_{23}	S_{33}
Thickness change (– %)	0.003	1.82	2.00	0.93	5.56	9.09	2.77	6.42	7.27
Diameter change (– %)	0.002	0.60	0.80	0.40	1.00	1.40	0.60	1.00	1.60

In group (A), it is noticed that the percentages change of the thickness, gradually increases as the applied pressure decreased. The same results observed in groups (B and C). In groups (D, E and F), the percentages change in thickness, increases with decreases in the firing temperature.

In groups (A, B and C), it is seen that the lowest percentage changes in diameter occurs with highest applied pressure. While, in groups (D, E and F), the lowest percentage change in diameter occurs with highest applied firing temperature.

Therefore, both higher applied pressure and temperature lead to lowest change in a diameter.

Water absorption

The data obtained from the water absorption test are shown on Table 14.7.

Table 14.7: Water Absorption

Group	*A*			*B*			*C*		
Specimen	S_{11}	S_{12}	S_{13}	S_{21}	S_{22}	S_{23}	S_{31}	S_{32}	S_{33}
Water absorption %	0.36	0.47	0.35	0.71	0.74	0.76	4.92	5.44	6.04
Group	*D*			*E*			*F*		
Specimen	S_{11}	S_{21}	S_{31}	S_{12}	S_{22}	S_{32}	S_{13}	S_{23}	S_{33}
Water absorption %	0.36	0.71	4.92	0.47	0.74	5.44	0.35	0.76	6.04

In groups (A, B and C), specimens S_{11} registered the lowest water absorption (0.36 per cent), and generally it is seen that water absorption decreases gradually as the applied pressure is increased.

In groups (D, E and F), the water absorption decreased gradually as the firing temperatures is increased.

The standers of water absorption percentage of General Hard Porcelain fall in the range (0.2 up to 2 per cent) (Singer and Singer, 1979). Thus, it is noticed in these results that specimens (S_{3x}) which fired at 1100 °C, have water absorption values out of the stander range, irrespective of the applied pressure. All other specimens fall well within the stander range.

Thus, the firing temperature and the applied pressure are effective factors on the water absorption, but firing temperature is the more effective factor on water absorption. This is probably due to the formation, at $T < 1100$ °C, of strong glass bonding which block micro spaces between porcelain crystalline phases.

Compressive Strength

The specimens were compressed until they broken down. The compressive loads were converted to compressive or crushing strength according to equation (2); the results obtained are presented in Table 14.8.

Table 14.8: Compressive Strength

Specimen	S_{11}	S_{21}	S_{31}	S_{12}	S_{22}	S_{32}	S_{13}	S_{23}	S_{33}
Compressive strength (MPa)	8.15	8.40	8.65	8.15	8.91	9.16	18.33	12.73	10.18

From Table 14.8, in groups (A), it is seen that specimen S_{13} has the highest compressive strength (8.65 MPa), and that the compressive strength, decrease gradually as the compaction pressure is increased. The same trend occurs in group (B). But it noticed that, in group (C), the compressive strength increase gradually as the applied pressure increase. Hence, it may inferred that the specimens of group (A and B), were fired well enough to form mature and strong glass bonding, but they have much glaze, and for this reason they gave low compressive strength in comparison with group (C).

In group (D) it is noticed that as the firing temperature is increased compressive strength decreased. The same is noticed in groups (E and F).

X-Ray Diffraction Data

It is observed that the diffraction pattern from (S_{31}), which has been heated at 1100 °C, shows the start of mullite formation (week peaks) Figure 14.3. (S_{31}), but contains unidentified peaks which may be due to unknown intermediate phases. The patterns obtained from (S_{11} and S_{21}), Figures 14.1 and 14.2 respectively, which heated at 1200 and 1150 °C respectively, indicate the presence of only silica and mullite, the other initial component have disappeared. It may be inferred that these

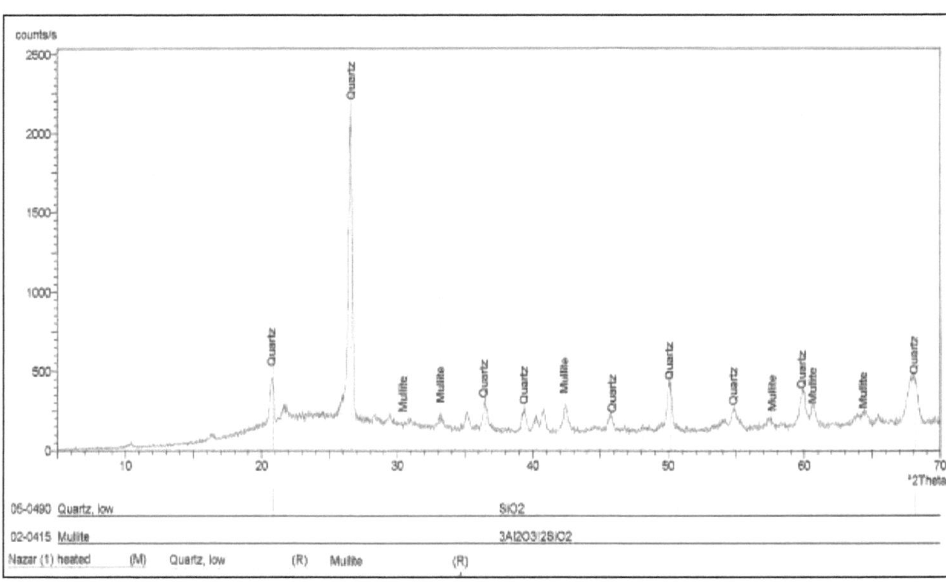

Figure 14.1: X-ray Diffraction Pattern of Specimens Pressed at 170 MPa and fired at 1200°C, (S_{11})

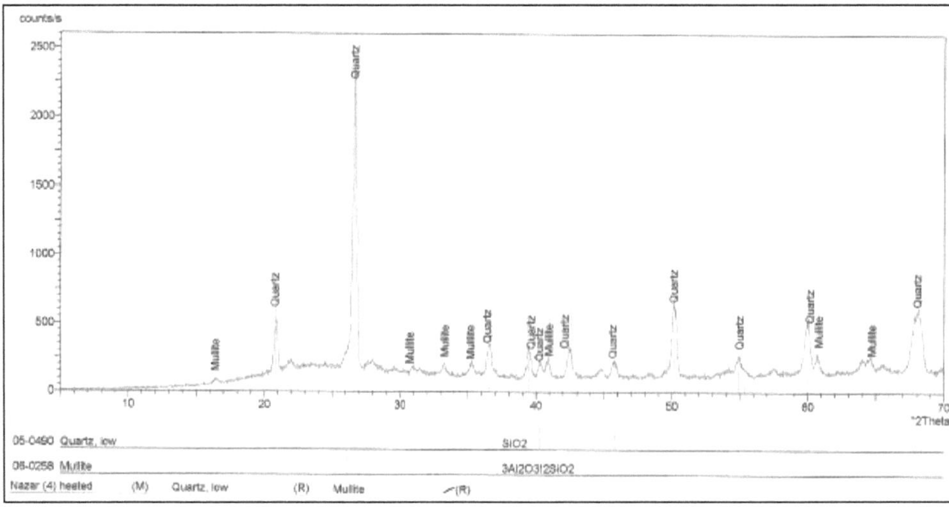

Figure 14.2: X-ray Diffraction Pattern of Specimens Pressed at 170 MPa and Fired at 1150°C, (S_{21})

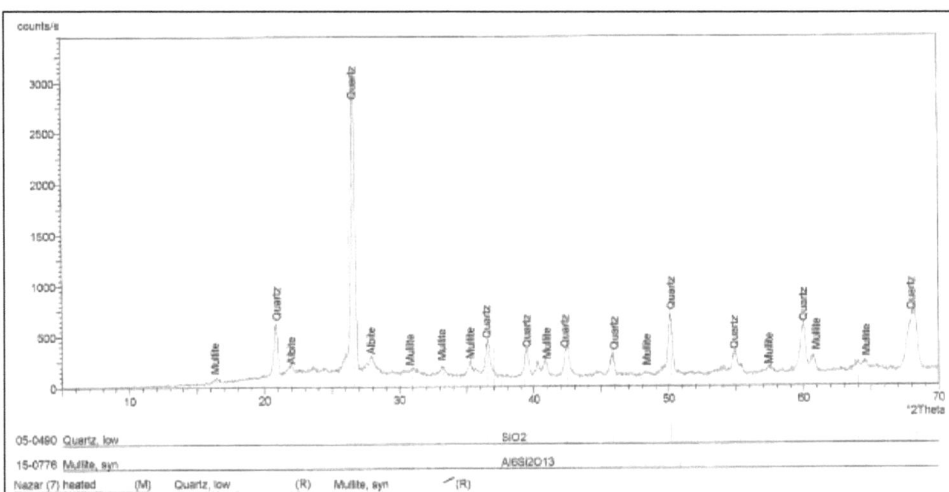

Figure 14.3: X-ray Diffraction Pattern of Specimens Pressed at 170 MPa and Fired at 1100°C, (S_{31})

components disintegrate on heating at (T < 1150 °C) giving rise to the formation of mullite.

The spectrum on powder mix, Figure 14.4, a mixture of raw materials, confirm that the dry powder consists only of quartz, kaolin, microcline (potash-feldspar) and talk.

The spectrum on Figure 14.1 (S_{11}), the specimen heated at 1200 °C, compared with that of the spectrum on powder mixed, Figure 14.4, indicates that important

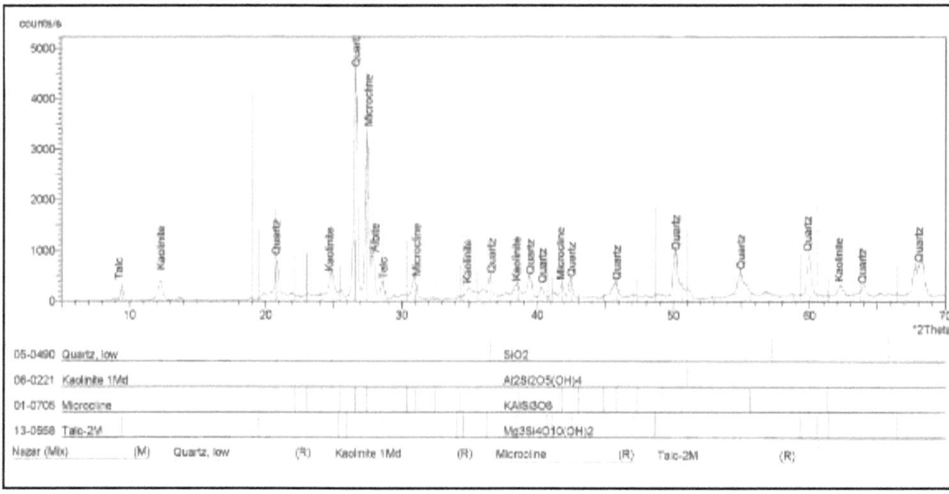

Figure 14.4: X-ray Diffraction Pattern of Unfired Blended of Raw Materials

changes have occurred. The original components (quartz, kaolin, microcline (potash-feldspar) and talk) have disappeared and have been replaced by mullite.

Generally, it is observed that the dry powder has a high ratio of quartz as all its constituents such as flint, potash feldspar and kaolin have high ratios of quartz in their formations which may not be completely converted to mullite.

In vitrified ceramics, adherence and strength are due to the formation of glass bonds (Kingery and Francois, 1967). The diffraction data obtained, shows that mullite is formed on heating the specified formulation. Aluminum atoms present in the mullite

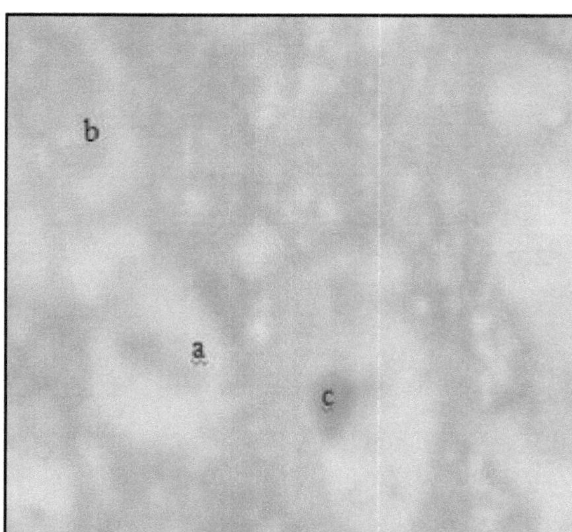

a: Residual quartz
b: Mullite
c: Iron impurity

Figure 14.5: Optical Microscopic Structure of Specimens Pressed at 170 MPa and Fired at 1200°C, (S11)

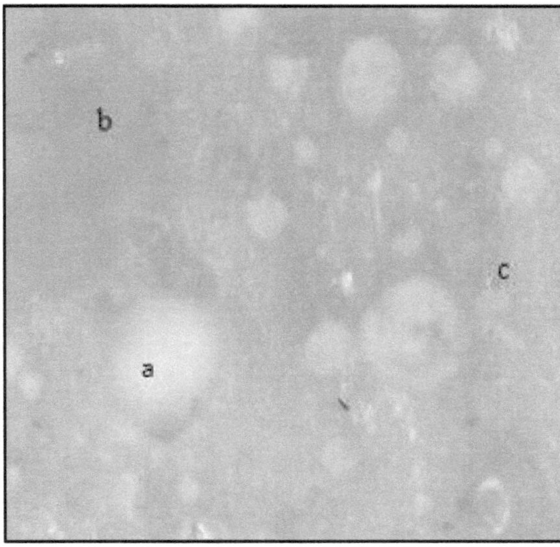

a: Residual quartz
b: Mullite
c: Iron impurity

Figure 14.6: Optical Microscopic Structure of Specimens Pressed at 140 MPa and Fired at 1200°C, (S12)

Figure 14.7: Optical Microscopic Structure of Specimens Pressed at 110 MPa and Fired at 1200°C, (S13)

formula ($3Al_2O_3.2SiO_2$), come from the kaolin and the potash feldspar. Probably the components disintegrate even at 1100 °C, fluxed by K_2O in feldspar and MgO in talc. This releases a large amount of silica and much smaller amount of aluminum. The mullite forms in limited quantity, but enough to show in diffraction.

Figure 14.8: Optical Microscopic Structure of Specimens Pressed at 170 MPa and Fired at 1150°C, (S21)

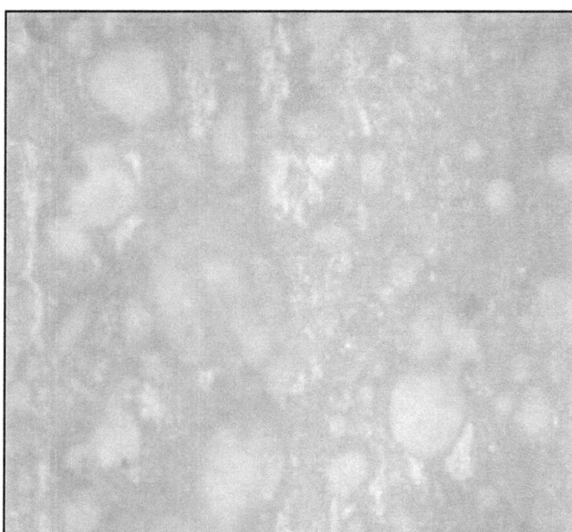

Figure 14.9: Optical Microscopic Structure of Specimens Pressed at 140 MPa and Fired at 1150°C, (S22)

The Microstructure

In group (A) Figures 14.5–14.7 which is (S_{11}, S_{12} and S_{13}) respectively, the formation of crystals have been noticed and observed, but crystals that formed in Figure 14.5 were more and have closer spaces between crystals than the rest. So, the inferred conclusions of this group are that the crystals have formed and shown clearly in all

Figure 14.10: Optical Microscopic Structure of Specimens Pressed at 110 MPa and Fired at 1150°C, (S23)

Figure 14.11: Optical Microscopic Structure of Specimens Pressed at 170 MPa and Fired at 1100°C, (S21)

Figures, also the higher applied pressure, the closer are the space between crystals, which means dense body.

In group (B), Figures 14.8–14.10, which is (S_{21}, S_{22} and S_{23}), the same observations as in group (A) are made.

In group (C), Figures 14.11–14.13, which is (S_{31}, S_{32} and S_{33}) respectively, the structure is fuzzy, no crystals are observed as clearly as in groups (A and B). A

Figure 14.12: Optical Microscopic Structure of Specimens Pressed at 140 MPa and Fired at 1100°C, (S32)

Figure 14.13: Optical Microscopic Structure of Specimens Pressed at 110 MPa and Fired at 1100°C, (S33)

plausible explanation is that the lower temperature of 1100°C did not enable the fusion of feldspar (m.p. = 1150°C). It seems that the fluxing action is needed to break the strong bonds of the kaolin structure, which is the main supplier of alumina for the formation of mullite.

Generally, brown spots and stains have been noticed in most figures, which are interpreted as due to Iron-phases.

Conclusions

The data obtained on porcelain point to the following conclusions:

The porcelain can be produced from a dry powder by dry pressing at pressure ($P \geq 110$ MPa).

An optimum firing temperature, in this work, is 1200 °C.

S_{11} records the highest dielectric strength; it breaks down at 17.4 kV. This indicates that, high temperature, at least up to 1200 °C, and high pressure, up to 170 MPa are needed to obtain high dielectric strength.

Dry pressed specimens showed good dimensional stability after firing at high temperature; only very slightly changes in the thickness and diameter, (-0.003 to -9.09) and (- 0.002 to -1.60) respectively have been recorded for S_{11} and S_{12}, both were processed at 1200 °C.

The specimens that have been pressed at the high pressure (170 and 140 MPa) and fired at the high temperatures (1200 and 1150 °C), exhibited the lowest water absorption capacity.

The compressive strength of the porcelain products is ranged between (8.15 to 18.33 MPa).

The mechanical properties of the porcelain products, *i.e.* compressive strength, are found to be strongly influenced by the compaction pressure, and the firing temperature.

The microstructure of these ceramics is shown to be predominantly crystalline, made of a mullite matrix and a silica phase. Strong evidence is found that the binding mechanism in these products is achieved by a mullite phase rather than by glassy bonds, which is the accepted mechanism in vitreous ceramics.

References

1. Griffin, E.A. and McMillin, S., 1995. *Ceramic Processing and Sintering*, Vol. 6. Marcel Dekker, New York.
2. King, A.G., 2009. *Ceramic Technology and Processing*, Norwish, NY; Noyes Publication, pp. 209–211.
3. Kingery, W.D., 1963. *Ceramic Fabrication Process*, 3rd Printing.
4. Kingery, W.D. and Francois, B., 1967. The sintering of crystalline oxides. I. Interactions between grain boundaries and pores. In: *Sintering and Related Phenomena*, (Eds.) G.C. Kuczynski, N.A. Hooton, C.F. Gibbon. Gordon and Breach, New York, p. 471.
5. Liebermann, J., 2001. Avoiding quartz in alumina porcelain for high-voltage insulators. *Am. Ceramic. Soc., Bull.* 80(6): 37–42, (7): 43–48.
6. Lucas, J.R., 2001. *High Voltage Engineering.*
7. Luikov, A., 1936. Moisture content gradients in the drying of clay. *Brit.*
8. Marinescu, I.D., Tonshoff, H.K. and Inasaki, I. (Eds.), 1999. *Handbook of Ceramic Grinding and Polishing*. William Andrew Publishing, Norwich, NY, USA.

9. Norton, F.H., 1952. *Elements of Ceramics.* Addition-Wesley, Cambridge 42, Mass, USA.
10. Rahman, M.N., 2003. *Ceramic Processing and Sintering,* 2nd Edition. New York, NY: Marcel Dekker, Inc., pp. 328–424.
11. Ryan, W., 1978. Fric, C. Chem., *Properties of Ceramic Raw Materials.*
12. Salih, S.A., 2005. Characteristics and potential industrial uses of Sudanese kaolin deposits. *PhD Thesis.*
13. Singer, F. and Singer, S.S., 1979. *Industrial Ceramics.*
14. Thurnauer, H., 1958. Controls required and problems encountered in production dry pressing.
15. Vincenzin, P.I. (Ed.), 1991. *Fundamentals of Ceramic Engineering.* Elsevier Science Publishers, Ltd., New York.

Chapter 15

Alloying of Steels with Boron Using Laser Techniques

*Khalil Azimeh**

Professor in the Department of Design and Production Engineering, Faculty of Mechanical and Electrical Engineering Damascus University P.O. Box: 31262, Damascus, Syria

ABSTRACT

The technology of the surface alloying using laser is the most diverse methods of surface treatment in terms of usage and the multiplicity of parts and components of machinery that could be addressed. Group is characterized of the laser alloying layer than the diffuse layer, which we get treated by thermo chemical method as a homogeneous chemically. The study discussed of a region of the laser alloying and related phenomenon's, and in the search used the boron carbide, chromium and boron powders as alloying elements for tool steel. Micro-hardness change was discussed in the laser alloying layer and study of microstructure change, as it increased the rate of micro-hardness 4-5 times. The microscopic structure of the laser alloying layer structure of the form of granules and fine-grained changes from ridge to form a multi-surfaces. Results showed an increase in wear resistance of layer after the alloy surface 3-8 times.

Keywords: *Tool steel, Laser alloying, Surface treatment, Wear.*

Introduction

Laser heat treatment because of its specificity allows to obtain high mechanical properties of surface-hardened layer. The level of properties achieved by this treatment, the chemical composition of the material details. For a number of steels not undergoing

---

* E-mail: khazimeh@scs-net.org

phase transformations during heating and cooling systems, laser treatment is generally ineffective [1,2].

A more promising method in terms of widening the range of hardened parts is the technology of laser surface alloying. The essence of the process consists of introducing into the zone of laser influence of alloying components. This results in mixing of additives with the molten core material. This results in a surface layer with a modified chemical composition. Varying the composition of the alloying components can be fairly wide range, changing the physical and mechanical properties of the surface, even on materials, heat treatment which does not give practical results [1].

The Mechanism of Forming a Zone of Doping

Laser doping is most often carried out by applying to the part surface coating containing the alloying elements, followed by melting the surface layers. The amount of absorbed energy of the material is determined by optical and thermal properties of the surface of base material and coating thickness. In laser doping, the energy of radiation of optical quantum generator (laser) is partially absorbed by the surface (E), is lost due to reflection from the coating ($E_{p.o.}$) and is absorbed by daubing including energy, do not come down to the treated material ($E_{obm.}$) (Figure 15.1):

$$E = E_{p.} + E_{p.o.} + E_{obm.}$$

This suma ($E_{p.} + E_{obm.}$) represents the "useful" part of the energy pulse, which is involved in the formation zone doping [3]. To predict the geometrical parameters of zones of laser alloying,it is necessary to know the threshold pulse power absorbed by

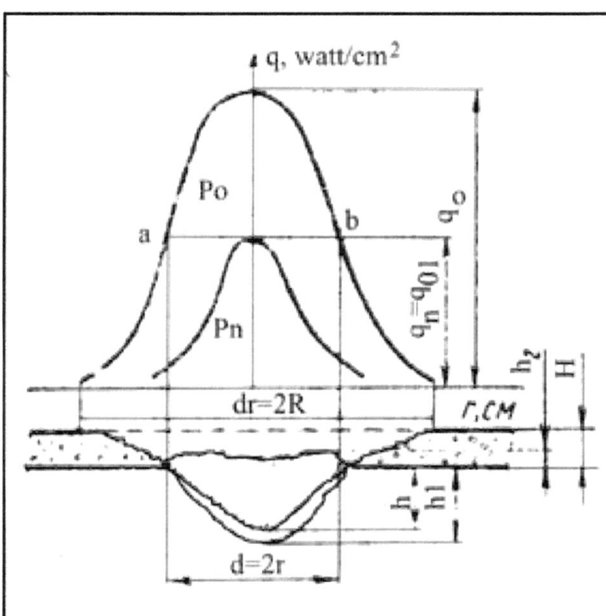

Figure 15.1; Schematic Diagram of Laser Interaction with the Metal Surface Layer with Coating Materials

the slip layer coating (P), and pulse power, (Pu) used in the formation of the zone of hardening. When exposed to normal-circular laser beam with an output pulse (Pu), the maximum power density in the center of the beam qo and the diameter of the focal spot of d on the metal surface covered by daubing including saturating element, the thickness H, a zone of doping, which has the following geometrical characteristics: diameter d, deep zone doping h, the depth of the melting zone h1. Diameter of doping is always smaller than the diameter of the laser beam, due to uneven distribution of power density over the cross section of the light flux having a maximum value at the center (qo) and decreases with distance from the center of the Gaussian as shown in Figure 15.1.

Difference $P_{l.l'}=(P_u - P_n)$is a useful power to be used for doping. In accordance with the calculations of AE:

$$P_{ll} = \frac{2\pi\lambda r(T_m - T_0)}{arctng\left(h_1/_r\right)}$$

where,

λ: Thermal conductivity of metal

r: Radius of the laser beam,

h_1: The depth of fusing,

T_m: Temperature of the solid phase at the interface between solid and liquid phases,

T_o: Temperature of the metal at a great distance from the metal-liquid.

The process of laser doping consists of the following steps [4]:

1. Heating of alloy composition to achieve in the contact zone with a base temperature of melting of the material.
2. Education of the liquid phase and phase mixing in a zone of molten bath.
3. Crystallization of liquid phase.

Upon reaching the border,"cover - sublayer" melting temperature (about 1400°C) is formed by liquid phase and the saturation of the surface layer of alloying element is intensified. At the first stage,doping occurs by diffusion in the liquid phase, in the second phase, another mechanism of doping - convective mixing phases. Established [3,5] that mass transfer in the molten bath flows from the center of the laser spot to the edges due to the gradient of surface tension caused by the temperature difference between the center and edges of the molten bath (Figure 15.2). In this case, the velocity of surface flow may reach 3 m/sec. The motion of the internal flow proceeds due to the forces of viscous friction at speeds of an order of smaller magnitude. As a result, circulation of the flow, changing the pressure of moving, flow is balanced by surface tension at the curved surface of the melt.

Qualitative characteristics of the doped layer is determined by the duration of stages. In the first stage of laser alloying is a significant oxidation coating, which

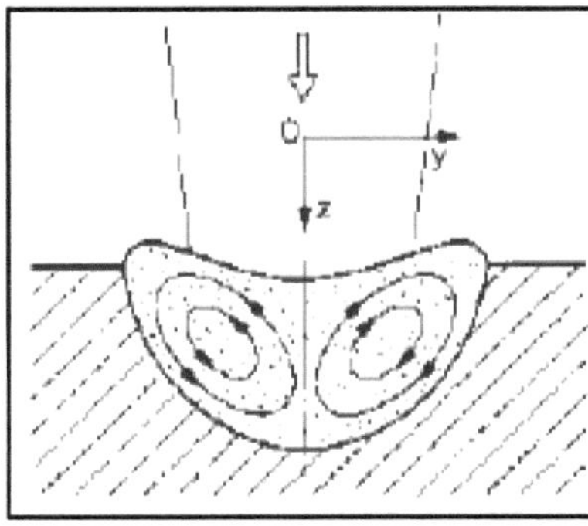

Figure 15.2: Schema Transport Mechanism of the Alloying Element within Melting Area

determines the basic characteristic of the quality layer at this stage - the degree of burnout. In the second stage, flow processes to ensure the mass transfer of alloying constituents in the molten bath. Quality coverage at this stage is determined by the degree of uniformity of the distribution of alloying components in it, and consequently, uniformity fizko mechanical properties. The quality of the surface layer increases with the reduction of the duration of the first stage (reduction burning) and increased duration of the second (ie, the degree of mixing of alloying elements). To do this; we must strive to ensure that the border "alloying layer - substrate quickly reached the eutectic melting temperature, and there was a formation of the liquid bath, which is heavily influenced by the height and pitch of the original surface asperities. The duration of the third stage of processing identifies such characteristics of the doped layer as the dispersed structure and residual stresses in it. With the decrease in the duration of this phase, decreases the dispersion of the structure, and increase tension in the surface layer, ie, an opportunity to cracking. In addition, the duration of phase determines the phase transformations during crystallization of the melt.

In [6,7] found that in the zones of laser alloying from the liquid state due to temperature and concentration gradients, occur high degree of disequilibrium of the liquid phase, the bifurcation does not melt stability and transition of laminar to turbulent fluid flow. This occurs when the vortices provide for different levels of scale energy dissipation, the abnormal flow of mass transfer and dispersion of the growing crystals, ie nanostructured effects. Thus, laser treatment of liquid-phase coating with the introduction of these dispersed particles of compounds of different elements makes it possible to obtain nanocomposite materials with desired composition, structure and properties. Great influence on the properties of laser alloyed surface layers of the process has virtually instantaneous energy absorption during laser processing, which conducts to the generation of stress waves, which accelerate the mass transfer of alloying elements from the coating in the base metal and high local plastic deformation of the microvolume of the metal.

Materials and Methods

Given the appointment of steel U10 (U.S. W1-9, Japan SK4- Table 15.1) for manufacturing the cutting tool and the need to improve the durability of the tool by increasing the hardness of the cutting edges, consider the results of doping with boron. The process of laser alloying were carried out on the air for the installation of pulse action «Kvant-15». Laser alloying was carried out at a power density q = 250 v/mm^2, velocity v = 4, 6, 10 mm/s and the thickness of the coating h = 0,2 mm (Table 15.2).

Table 15.1: Chemical Composition of Steel Type U10

C	*Si*	*Mn*	*S*	*P*	*Cr*	*Ni*	*Cu*
0.9-1.04	0.17-0.33	0.17-0.3	0.028	0.03	0.2	0.25	0.25

Table 15.2: The Condition of Laser Process Parameters

q = 250 watt/mm^2	Power density
V = 4, 6, 10 mm/sec	Velocity
0.5 d_n = 0.3 mm	Overlaid tracks of laser doping
d_n = 0.6 mm	Diameter spot from a laser beam
4 * 10^{-3} sec	Pulse duration

Discussion and Results

A distinctive feature of the coating formed by laser alloying of diffusion coatings obtained by methods of chemical-thermal processing, chemical homogeneity and lack of bedding. Due to convective mixing of alloying, elements are evenly distributed throughout the area of doping, there is no transition from a phase with a higher concentration of alloying elements to the phases with less. Large cooling rate of molten lead to the formation of supersaturated metastable structures of high dispersion and hardness. Currently, there are quite a lot of experimental data on laser alloying of metallic alloys of various alloying elements. The general regularities of structure formation during laser doping. Sufficiently studied in detail the question of laser alloying non-metallic components. Mechanism of surface hardening of metal alloys by laser alloying with carbon, boron and nitrogen, as in conventional thermochemical treatment is based on a synthesis of the structure of the alloy in the melting zone of chemical compounds with high hardness and high heat resistance - carbides, nitrides, and borides [8,9]. According to [9], the saturation of the molten zone Armco - iron, low or of medium carbon steel of the fluid leads to the formation of its structure of high-carbon martensite, residual austenite, and carbides when carbon concentration is high. Microhardness of the alloy surface increases from 1500-2000 to 9000-14000 MPa. In the laser nitriding steels observed nitrous martensite, iron nitride (Fe3N, Fe4N). Microhardness of the surface increases up to 5500-6500 MPa. The largest contribution to the hardening of the solid-making mechanisms of solution hardening nitrides of alloying elements [9]. To achieve maximum hardness of the ferrite matrix alloy V, Cr, and Mo is necessary to conduct the nitriding followed by aging or increasing the duration of nitriding for the most complete selection of nitride phases. The exceptions are cover-doped Al, the maximum hardening γ, - phase (Fe, Al) 4N during nitriding [9,10]. The structure and microhardness of coatings produced by laser fusion of boron-containing coatings on steel surfaces, depend on the treatment regime. Micro-hardness in the structure, which consists of ferrite and boride eutectic is 6000 - 12000 MPa [11]. Saturation of the surface of boron by reducing the depth of melting or fusing of multiple application and coating gives rise to the structure of a large number of borides (FeB, Fe2B) and an increase in microhardness up to 14000 - 21000 MPa. Moreover, the high hardness of the coating remains on heating to 600 ° C [12]. When the doping iron alloys with refractory (W, V, Ti, Ta, etc.) or carbide (Cr, Mn, Mo) metal-doped coating microhardness increased. In alloying, elements do not form carbides (Ni, Cu, Al, etc.), the microhardness decreases. In addition to increasing the microhardness of doped with chromium; significantly increases the corrosion resistance of the surface [13].

The structure and microhardness of the doped layer depends on many factors. It plays an important role of the alloying additives, the regime of laser treatment, the method of feeding of the alloying material in the melting zone.

Crystallization of molten bath occurs at very high speeds due to having a major amount of heat into the metal framework after the termination of the laser pulse. At high cooling rates observed in pulsed laser processing, the microstructure of the molten zone not found a coarse-grained structure. Figure 15.3 shows a view of the

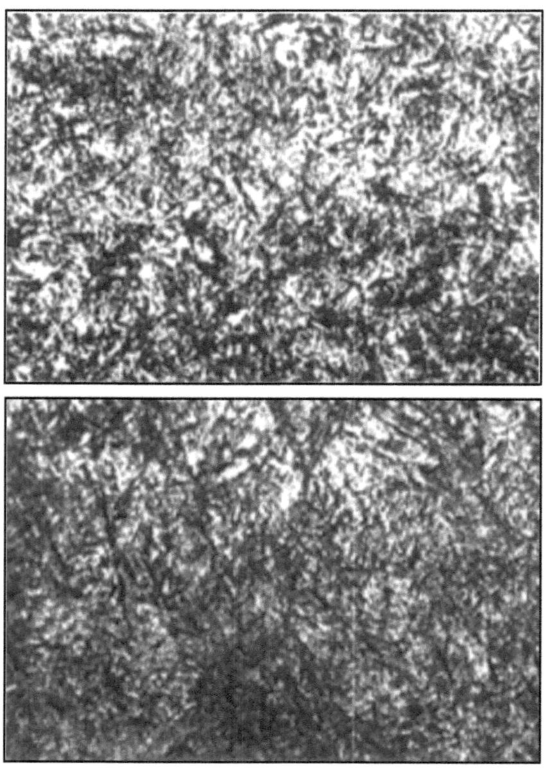

Figure 15.3: The Structure of Steel U10 after Laser Doping with Boron X400

(a) In a bath of doping, and (b) The transition zone between the layer and the metal base

doping bath boron doping zone has fine-grained structure and the grain shape changes from prolate to polyhedral.

Figure 14.4 shows the distribution of microhardness after boron doping for different speeds. With increasing velocity of the microhardness in the tub, doping increases. Such dependence is explained by the fact that the distribution of alloying elements occurs in a smaller volume of molten metal, which causes an increase in tension structure in the tub doping. With decreasing velocity of the beam to 2 mm/s is the fall in the microhardness due to burnup of boron from the surface.

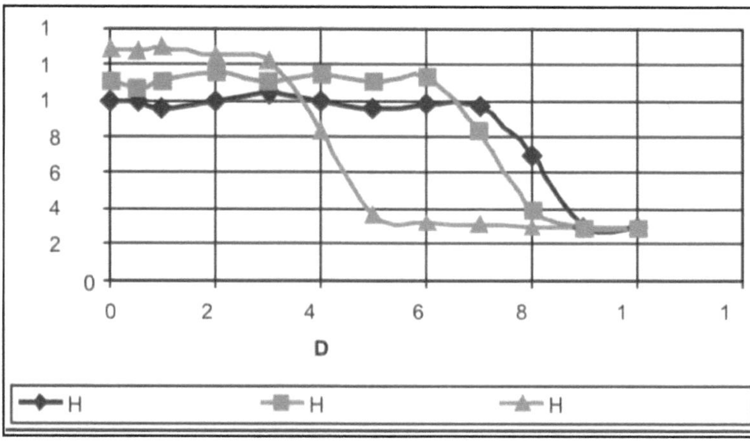

Figure 14.4: Microhardness Distribution Steel U10 after Laser Alloying with Boron

When you change the speed of movement, it varies the geometrical parameters (depth and width). So with increasing velocity of the bath, depth of the doped layer decreases (Figure 14.5). This character of the curve can be explained by the fact that the change in velocity leads to a change in the energy deposition.

The analysis showed that the laser doping significantly increases the microhardness of the steel surface to - 14000 MPa, compared with the hardness of the

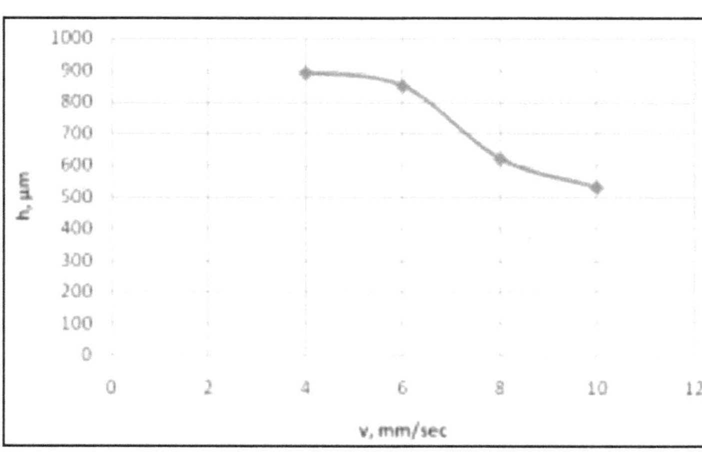

Figure 14.5: Effect of Velocity on the Depth of the Laser Doping with Boron Steel U10

substrate (the initial hardness - 3000 MPa). The high microhardness values associated with the characteristics of the microstructure are determined by a high content of carbides and borides, high dispersion and supersaturated solid solutions is contained within the molten bath. Solid solutions have a considerable degree of supersaturation due to the fact that the iron atoms are replaced by atoms of chromium in the crystal lattice sites, as well as the introduction of boron and carbon in the pores of the lattices of solid solutions.

The tests of wear resistance of the surface of steel U10 after laser doping with boron are estimates and shows the level of improving performance. Hardened by laser surface alloying of steel even at higher specific load remains very low rate of wear compared to standard hardened by heat treatment condition.

Consideration of technological methods can be used to ensure and improve the contact strength of functional surfaces as in the restoration of worn parts (with a limited value of depreciation), and in hardening of the newly produced.

Conclusions

1. The doping bath boron doping zone has fine-grained structure and the grain shape changes from prolate to polyhedral.
2. With increasing velocity of the microhardness in the tub, doping increases.
3. With increasing velocity of the bath, depth of the doped layer decreases.
4. The surface of steel U10 after laser doping with boron are estimates and shows the level of improving performance.

References

1. Devoino, O., Golovachenco, A. and Kardapolova, M., 1990. Improving the physical and mechanical properties of friction surface laser treatment. *Minck,* 40 p.
2. Grigoryans, A., Shiganov, I. and Misurov, A., 2008. *Technological Processes of Laser Processing.* Moscow, 664 p.
3. Lakhtin, U. and Kogan, Y., 1990. Surface alloying of metals and alloys by laser heating. *Moscow,* 56 p.
4. Cpiridonov, N., Kobyakov, O. and Kupriynov, I., 1988. Plasma and laser methods hardening of machine parts. *Minck,* 155 p.
5. Sarkar, S., Mohan Raj, P., Chakraborty, S., Phanikumark, G. andDutta, P., 2003. Transport phenomena in laser surface alloying. *Journal of Material Science,* 38: 155–164.
6. Brover, G. and Krienin, C., 2008. Structural features of laser alloying of steel coatings of defferent composition. *Modern Problems of Science and Education.* Nauka N6.
7. Grigoryans, A., Misurov, A. and Chjan, S., 2007. Forming layers deposited using laser pulse-periodic radiation. *Welding Factory,* N8. pp. 18–21.

8. Arutunyn, R. 1988. Sated surface layer of iron with carbon laser alloying of liquid. *Surface, Physics, Chemistry, Mechanics.* ¹5. Ñ.127–129.
9. Novecov, V. and Latshiv, V. 2000. Modification and strengthening of the rubbing surfaces laser treatment. *Ivanovo*, 119p.
10. Chodina, O., Petrova, L. and Berekovaskaya, T., 2002. Hardening mechanism of iron in laser alloying and nitriding. *Metallurgy and Heat Treatment of Metals*, N4: 20–26.
11. Devoino, O., 2000. The influence of the laser alloying technological parameters on linear characteristics of the strengthened zonts. *Tools– 200 : Prjc. Int. Conf. Trenchin.*
12. Govorov, I., 2006. Increase the contact strength of functional surfaces of steel products during laser processing. *Bulletin of Bryansk State Technical University*, N2(10).
13. El Kadiri, H., 2008. Phase transformations in low-alloy steel laser deposits. *Materials Science and Engineering*, A494: 10–20.

Chapter 16

Euphorbia Lagascae: A Plant Resource for the Production of Polymers

*L.Y. Mwaikambo**

Department of Mechanical and Industrial Engineering, University of Dar Es salaam, P.O. Box 35131, Dar Es salaam, Tanzania

ABSTRACT

Euphorbia (Euphorbia *lagascae*) which is an oil crop consisting of a large quantity of vernonia acid with high levels of expoxidated triglyceride had undergone chemical modification in its structure to obtain hydroxyl groups. The hydroxylated chemical structure was crosslinked using methylene-p-phenyl diisocyanate to produce a thermoset euphorbia oil resin. The cross linking process was monitored *in situ* using the Attenuated Total Internal Reflectance Fourier Transform-Infrared spectroscopy and the thermogravimetric analysis techniques. The infrared technique was also used to monitor chemical groups of the unmodified euphorbia oil. A reduction of the absorption band around 3300 cm^{-1} which is due to the presence of hydroxyl groups was observed indicating the occurrence of the reaction between the hydroxylated euphorbia oil and methylene-p-phenyl diisocyanate. The thermogravimetric analysis technique recorded the reduction of the hydroxyl groups as the loss in mass of the reactants as a function of temperature. The two analytical techniques provided about the same cure time. Both, the infrared and thermogravimetric techniques elucidated useful cure characteristics of the euphorbia oil resin. The applications of more than one analytical method offer a complement of the results. The use of plant resources oils

* *Corresponding Author:* E-mail: lyrmwaikambo@udsm.ac.tz

provides a new and renewable source of polymers which can be used for making shoe soles and mattresses thus providing added value benefits to developing economies.

Keywords: *Euphorbia lagascae, Triglycerides, Hydroxylation, Cure, Polyurethane.*

Introduction

Non food oil crops have been found to consist of expoxidated triglyceride structure (Liu *et al.*, 1998, Edqvist, and Farbos, 2002). This discovery has lead to an intensive research to determine the potential of such oil crops as renewable feedstokes for the production of polymers. These plants are of different variety and the most common are *Euphorbia lagascae* and *vernonia* both of which consists of fatty acids of expoxleic acid (12, 13 epoxyoleic acid) and vernoleic acid respectively(Liu *et al.*, 1998, Edqvist, and Farbos,2002, Krewson and Scott 1996). It is reported that certain plants consists seeds that provide large quantities of epoxy bearing triglycerides. The successes in these researches to produce renewable raw materials for the polymer industry will revolutionise the dependence on fossil-origin precursors as feedstokes for the synthesis of polymers. *Euphorbia lagascae* has been found to consist of a large content of diepoxy followed by triepoxy and monoepoxy while vernonia oil consist large content of triepoxy followed by diepoxy and monoepoxy respectively. Oils with higher content of exirane and unsaturation should give polyols with higher functionality (Hu *et al.*, 2002) making them suitable for the production of epoxy polymers as well as polyurethane (Hayes, 1996). For instance, vernolic acid is an epoxy acid with application in paints, coatings and polyurethane polymers (Purdy and Craig, 1987). Reactions of epoxy consisting structures can undergo nuceophilic addition to an epoxide under acidic or basic catalysts environment, hydrolysis of an epoxide in the presence of an acid catalyst generates a glycol, reduction of an epoxide with lithium aluminium hydride and water generates an alcohol under basic conditions.

Generally, oils are converted into polyols through two steps, first by epoxidation of the double bonds followed by hydroxylation (see scheme 1). Euphorbia oils (*Euphorbai lagascae*) are triglycerides of fatty (C_{12}–C_{22}) acids, mainly consisting of mono-unsaturated fatty acids such as oleic acid and poly-unsaturated fatty acids such as linoleic and linolenic acids (Javni *et al.*, 2000). A schematic representation of the structure of epoxy bearing vernolic acid in euphorbia oil (Neff *et al.*, 1993, Purdy and Craig, 1987) and its corresponding structures after epoxidation followed by hydroxylation is shown in Scheme 1.

Vegetable oils have a number of excellent properties, which could be utilised in producing valuable polymeric materials such as polyurethane. Polyurethane is a generic term used to classify polymers composed of synthetic linear macromolecules having in the chain recurring aliphatic groups joined by urethane groups which together comprise at least 85 per cent (by mass) of the chain (Beech at al. 1986). The structure of these polyurethanes can be tailor made to suit specific requirements by selecting the type of polyols and based on the hydroxyl group density (Javni at al. 2003, Monteavaro *et al.* (2005). Polyurethanes are produced via the reaction of polyisocyanate with polyols and water in the presence of catalysts (Badri *et al.*, 2001).

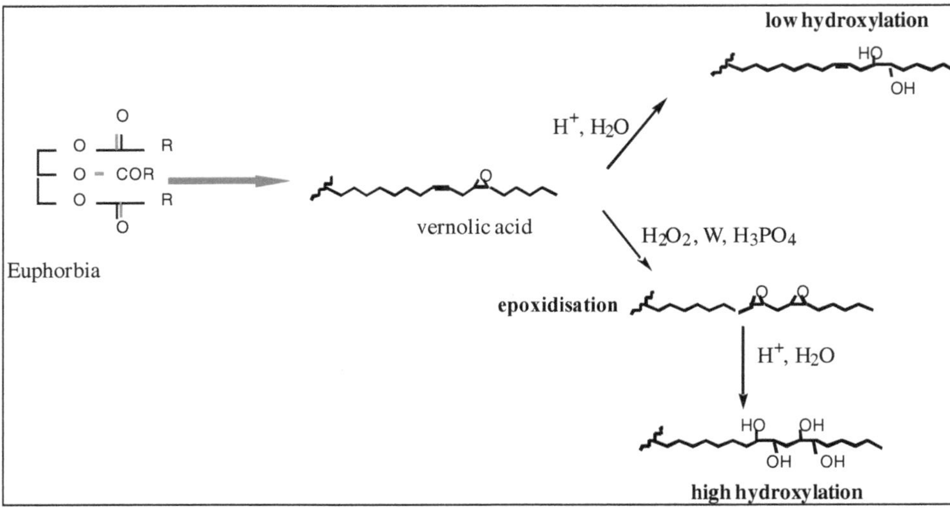

Scheme 1: Schematic Representation of the Euphorbia Triglyceride

The formation reaction of polyurethane polymers basically consists of two routes. The first route (1) involves the reaction of isocyanate with the polyol to form urethane and the second route (2) involves the reaction of isocyanate and water to form urea and carbon dioxide. Scheme 2 shows two (Reactions (1) and (2)) general reactions routes involved in the production of polyurethane polymer.

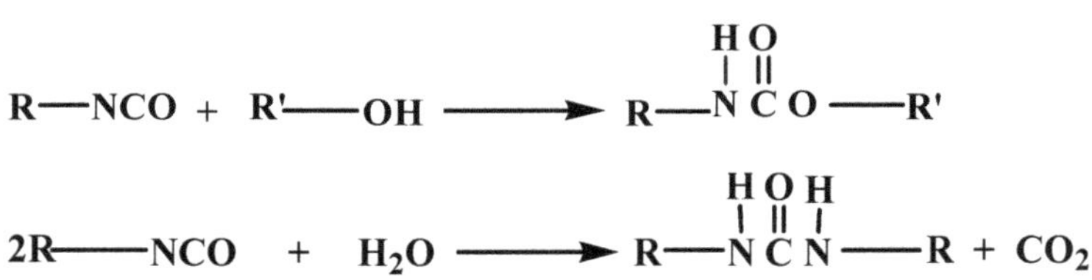

(1)

(2)

Scheme 2: The Two General Reactions Routes Involved in the Production of Polyurethane Polymer

The balance of the two reactions in Scheme 2 is quite critical to obtain the required physical properties (Javni 2000). The cure characteristics of most fossil-based synthetic polyurethane indicates complete resin cure temperature of around 50 - 60 °C within 45 - 60 minutes (Javni *et al.,* 2003, Javni *et al.,* 2000, Ravey and Pearce 1997). Currently, research work has shown that it is possible to produce monomers from plant-based oils as indicated in Scheme 1, which can be used for the synthesis of polymers (Filip *et al.,* 1992). However, whereas the cure characteristic of fossil-based resins has been well documented (Ravey and Pearce 1997, Chao *et al.,* 1998, Shibata *et al.,* 2002, Belousova 2002) those of plant-based polymers have not been thoroughly carried out. It has been found that the chemical structure of virgin euphorbia oil can be modified to obtain reactive groups such as hydroxyl groups (Hu *et al.,* 2002, Ren *et al.,* 2007) using inorganic catalysts and peroxides mixed with water. A general reaction is shown in Scheme 1. A detailed hydroxylation process of the oil will be reported in a future publication.

The introduction of hydroxyl groups on the euphorbia triglyceride is essential for polymer synthesis (Hu *et al.,* 2002, Ren *et al.,* 2007). In this study, a polymer was synthesised by reacting hydroxylated (polyol) euphorbia oil with methylene-p-phenyl diisocyanate (MDI). The chemical composition of the virgin euphorbia oil, hydroxylated euphorbia oil and MDI polymerised hydroxylated euphorbia oil were analysed using the Attenuated Total Internal Reflectance Fourier Transform-Infrared (ATIR FT-IR) technique. The thermal analysis of the virgin euphorbia oil, hydroxylated euphorbia oil and MDI polymerised hydroxylated euphorbia oil were also investigated using the Thermogravimetric Analysis (TGA). The resulting polymer has characteristics similar to conventional polyurethane reported by Ravey *et al.,* 1997 and Belousova 2002.

Materials and Methods

Materials and Equipment

Hydroxylated euphorbia oil was synthesized in the laboratory of the Department of Chemistry University of Warwick. Other chemicals used also included Methylene-p-phenyl diisocyanate (MDI) and toluene bought from Aldrich Chemicals. The equipment used in this study was Fourier Transform Infrared (FT-IR) and Thermogravimetry (TG).

Infrared Analysis

The Attenuated Total Internal Reflectance Fourier Transform-Infrared (ATIR FT-IR) model Nicolet was used to collect the spectra of *in situ* curing of EURE. Blobs of the mixture of hydroxylated euphorbia oil and methylene-p-phenyl diisocyanate were dissolved in toluene and the solution placed on the FT-IR bottom 'plate' and a head was lowered so that it just touched the solution. The whole set-up was then closed and the sample was scanned at room temperature (20 °C). The control on the preparation speed was essential in order not to lose significantly the initial cure of the resin. The spectra of the cross linking resin were collected and analysed.

Thermogravimetric Analysis

The reaction was studied by monitoring the loss in mass as a function of temperature of the resin. Pelkin Elmer Thermogravimetry (TG7) equipped with thermal analysis controller Thermal Analysis Controller TAC 7/DX was used. Samples of the mixture of hydroxylated euphorbia oil and methylene-p-phenyl diisocyanate solution weighing between 5 - 10 mg were placed in a crucible in the furnace chamber and heated in dynamic mode in nitrogen between 40 °C - 900 °C at the rate of 10 °C/min. The isothermal mode was operated at temperatures between 20 °C and 80 °C in steps of 5 °C. The collected thermograms were analysed.

Results and Discussion

Fourier Transform Infrared Analysis

Figure 16.1 shows the transmittance spectra of virgin euphorbia oil, hydroxylated euphorbia oil and MDI polymerised euphorbia oil euphorbia resin (EURE). It is observed that the virgin euphorbia oil shows no absorption bands above 3300 cm^{-1}

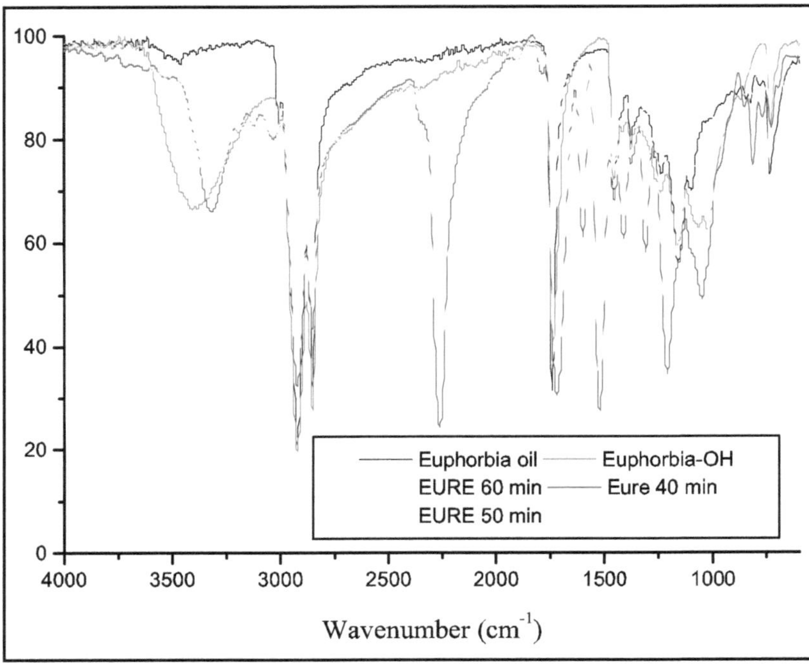

Figure 16.1: FT-IR Spectra of Unmodified Euphorbia Oil, Hydroxylated Euphorbia Oil and MDI Polymerised Hydroxylated Euphorbia Oil. The results show appearance of peaks following reaction between the oil and isocyanate group.

indicating the absence of hydroxyl groups and amine groups but shows an intensive absorption bands at 2852 cm^{-1} and 2915 cm^{-1} indicating the presence of C-H bond (Ravey *et al.,* 1997, Chao *et al.,* 1998). The absorption observed at 2274 cm^{-1} for the cross linked polyol is due to the presence of the isocyanate group present in MDI (green, blue and light blue colour spectra). Other absorptions of interest to this study are observed at 1740 cm^{-1} and near 1164 cm^{-1} representing the C=O and aliphatic ether chemical groups, respectively (Chao *et al.,* 1998). The presence of the carboxyl groups at 1740 cm^{-1} is of great interest in reactivity particularly in the cross-linking with cellulosics (Shibata *et al.,* 2002). Following hydroxylation a broad absorption is observed between 3300 - 3500 cm^{-1} indicating the presence of O-H groups (red colour spectrum). The reduction in the absorption bands at 2852 cm^{-1} and 2915 cm^{-1} for polymerised euphorbia oil compared to unmodified euphorbia oil shows that some of the C-H groups have participated in the grafting of the O-H groups. However, the reduction in the absorption band is not significant implying that the reaction was at an early stage. Similar results are observed at 1740 cm^{-1} and 1164 cm^{-1} band. The MDI cross-linked EURE shows a decrease in the width of the transmittance band shifting towards higher wavenumbers between 3300 - 3500 cm^{-1} which implies that some of the O-H groups have been used in the polymerisation reaction (Ravey *et al.,* 1997, Chao *et al.,* 1998). Similarly the reduction in the band intensity of the EURE at 1740 cm^{-1} (Figure 16.1) entails that some of the carboxyl groups (C=O) participates in the polymerisation reaction. Similar observations were reported by Chao, *et al.,* 1998 when carrying out a study on the synthesis and physical properties of fire retardant polyurethane ionomers using the FTIR technique. The aromatic double bond is observed at 1600 cm^{-1} while the amine (NHCO) is observed at 1525 cm^{-1} band, which is also characteristic of the benzene ring (Chao *et al.*, 1998).

It is worth noting that whereas the presence of C=O groups indicates that the – hydroxylated euphorbia would not require a catalyst; the presence of the NCO in the polymerised EURE indicates possible further reactions (cure) of the polyol to form polyurethane polymer. Since, the determination using the hydroxyl value is tedious and time consuming particularly in industry. The FT-IR method adopted in this study was considered to be convenient, quick and comparably accurate.

Two absorption bands representing the hydroxyl groups between 3300 - 3500 cm^{-1} and isocyanate groups at 2274 cm^{-1} were selected to monitor the polymerisation reaction. The ratio of the percentage absorption of isocyanate:hydroxyl groups (NCO:OH) were calculated and plotted in Figure 16.2 against reaction time by integrating area under the curve representing the O-H and isocyanate group bands, respectively. The integration of the bands provides a simple technique for quantitative analysis of each functional group. A characteristic decrease in the ratio of intensity of NCO:OH is observed (Figure 16.2). Similar analogies were propounded by Chao *et al.*, 1998 when studying the synthesis and physical properties of fire-resistant polyurethane ionomers and by Tsai, *et al.*, 2008 when analysing the infrared of moisture curable polyurethane resins. The method used in this work is viewed useful particularly in industry where quick and accurate results are needed for quick interventions.

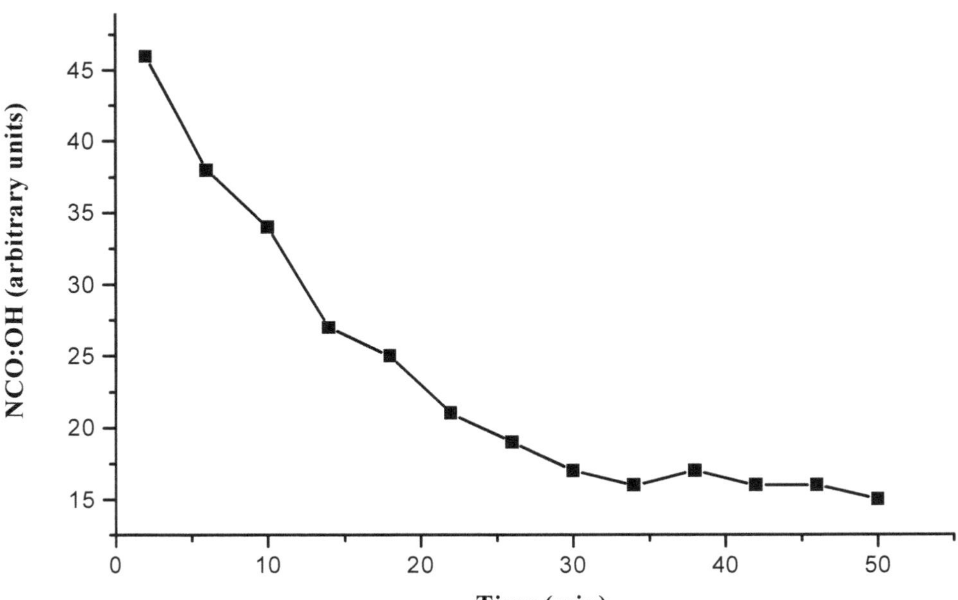

Figure 16.2: The Ratio of the Change of Isocyanate to Hydroxyl Groups with Respect to Reaction Time. The results show a decrease in the NCO:OH ratio with reaction time.

The decrease in the unreacted isocyanate is observed with an increase in the reaction time by studying the ratio of band NCO:OH illustrated in Figure 16.2. Close observation of Figure 16.2 provided average reaction time of between 45 minutes. It therefore implies that the optimum cure time of the euphorbia oil based resin is between 40 - 50 minutes.

TG Analysis

Figure 16.3 shows the rate of weight loss of the polymerisation of EURE in nitrogen at different temperatures. The maximum rates of mass loss were determined from Figure 16.3 for resins undergoing polymerisation in the thermogravimetry furnace for samples reacted at constant temperature. However, only representative curves are shown in Figure 16.3 while rates of mass loss of all the samples studied is plotted in Figure 16.4.

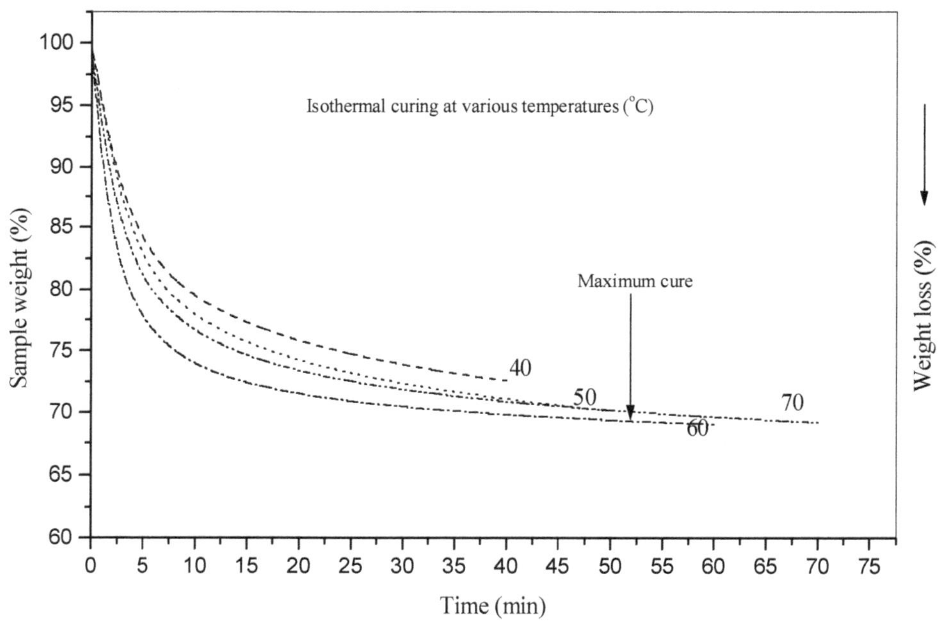

Figure 16.3: The Thermogravimetry Isothermal Cure of Various Temperatures. The results show reaction reaching completion with increase in temperature up to an optimum temperature and time indicated by an arrow.

Figure 16.3 shows curves exhibiting high initial rate loss at higher temperatures and that the optimum rate loss is attained around 45 minutes of reaction time and that the optimum temperature of cure is 60 °C.

Figure 16.4 is a plot of relative percentage cure after 60 minutes cure segments, exhibiting a fast resin cure within the first 25 minutes followed by gradual reaction process. It is observed that the optimum cure temperature occurs at 60 °C. Greenberg and Kamel 1977 used a similar approach to study the kinetics of anhydride formation in poly (acrylic acid) composites. Maximum weight loss occurs at the optimum curing temperature at 100 per cent degree of conversion.

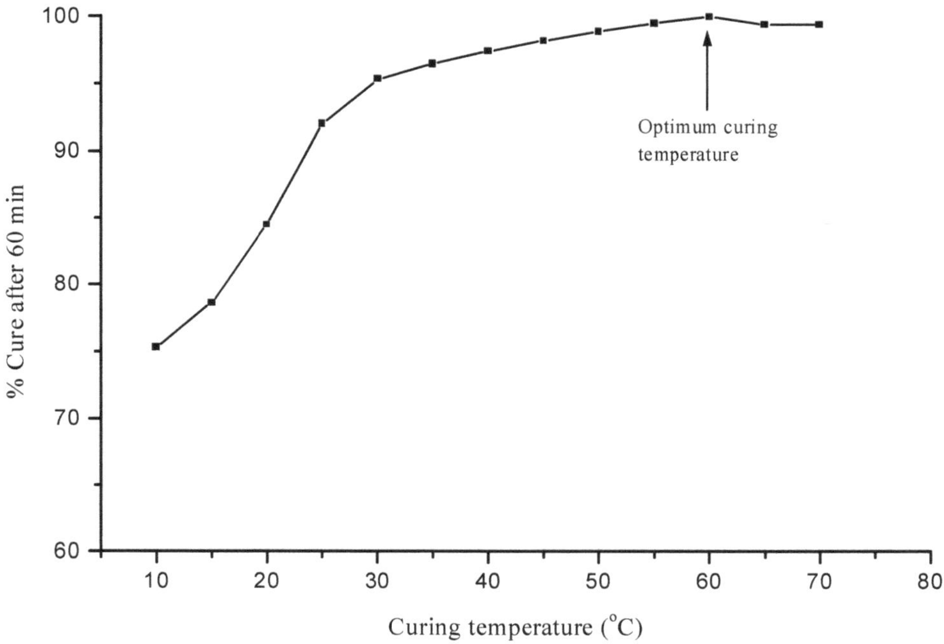

Figure 16.4: Isochronal Degree of Conversion of EURE against Cure Temperature. The results show an optimum cure temperature at 60°C.

Under ideal conditions, the weight loss is equivalent to the enthalpy loss by differential scanning calorimetry as a measure of the degree of conversion (Prime 1997). Maximum rate of conversion occurs at the highest gradient of the cure curve (Figure 16.4). It has been reported that the resin attain lowest viscosity at highest rate of conversion (Michaeli and Kamps 2007). In composite processing, the flowability of the resin is optimum at lowest resin viscosity (Michaeli and Kamps 2007). It is expected, therefore, that the resin's ability to effectively wet the reinforcement is expected to be effective at higher temperature of curing where it would exhibit lowest viscosity before the onset of cure. Thermoset exhibit low viscosity at high temperatures prior to cross linking (Michaeli and Kamps 2007). This allows easy flowability and in composites it facilitates fibre wetting critical for a coherent interface

Conclusions

In this study, it has been demonstrated that the FT-IR and TGA can be used to monitor the *in situ* cure characteristics of EURE. It has also been observed that FT-IR technique is more efficient in monitoring cure characteristics as it provides lower temperatures compared to TGA techniques. This implies that the application of heat may lead to degradation of reacting functional groups thus providing pseudo cross linking time. However, thermogramitric technique elucidate increasing reactivity of the isocyanate as the processing temperatures are increased. It is important to note that the cure analysis undertaken in this study covered both the initial rapid cross linking and the post cure stage of the resin. The delayed initial cure time reported using the TG method compared to FTIR value is due to the fact that the TG was run above room temperature. This implies that temperatures above 20 °C facilitate the cure activity of the MDI cross linked EURE reported in this work. A study of the cure characteristics is useful for the prediction of the temperature distribution and time for material processing. The use of Euphorbia *lagascae*, which is renewable as a feedstock for polymer production has economical and environmental benefit and providing prospects for increased agriculture and farmer income in East Africa.

Acknowledgements

The author wishes to acknowledge the support from The University of Dar es Salaam by allowing the researcher to carry out the work leading to the production of the published research findings.

References

Badri, K.H., Ahmad, S.H. and Zakaria, S.J., 2001. Production of a high-functionality RBD palm kernel oil based polyester polyol. *J. Applied Polymer Science*, 81: 384–389.

Beech, S.R., Farnfield, C.A., Whorton, P. and Wilkins, J.A. (Eds.), 1986. *Textile Terms and Definitions*, 8th Edn. Manara Printing Services, London.

Belousova, T., 2002 IR spectroscopic characteristics of polyacrylonitrile copolymer fibres. *Fiber Chemistry*, 34: 146–150.

Chao, D.Y., Kuo, W.J., Wang, N.H., Lee, C.L. and Yang, K.H., 1998. Study of the synthesis and physical properties of fire retardant polyurethane ionomers. *J. Applied Polymer Science*, 67: 19–26.

Filip, V., Zajic, J. and Smidrkal, J., 1992. Methanolysis of euphorbia oil triglycerides. *Revue Francaise des Corps Gras*, 39: 91–94.

Edqvist, J. and Farbos, I., 2002. Characterisation of germination-specific lipid transfer protein from *Euphorbia lagascae*. *Planta*, 215: 41–50.

Greenberg, A.R. and Kamel, I., 1977. Kinetics of anhydride formation in poly (acrylic-acid) and its effect on the properties of a PAA-Alumina composite *J. Polymer Science, Polymer Chemistry*, 15: 2137–2149.

Hayes, D.G. and Kleiman, R., 1996. Supercritical fluid chromatographic analysis of new crop seed oils and their reactions. *Journal of American Oil and Chemists Society*, 73(12): 1691–1697.

Hu, Y.H., Gao, Y., Wang, D.W. and Hu, C.P., 2002. Rigid polyurethane foam prepared from euphorbia oil based polyol. *J. Applied Polymer Science*, 84: 591–597.

Javni, I., Petrovic, Z.S., Guo, A. and Fuller, R., 2000. Thermal stability of polyurethanes based on vegetable oils. *J. Applied Polymer Science*, 77: 1723–1734.

Javni, I., Zhang, W. and Petrovic, Z.S., 2003. Effect of different isocyanates on the properties of soy-based polyurethanes. *J. Applied Polymer Science*, 88: 2912–2916.

Krewson, C.F. and Scott, W.E., 1996. *Euphorbia lagascae* Spreng.: An abundant source of epoxyoleic acid; seed extraction and oil composition. *Journal of American Oil Chemists' Society*, 43(3): 171–174.

Liu, L., Hammond, E.G. and Nikolau, B.J., 1998. *In vivo* studies of the biosynthesis of vernolic acid in seed of *Vernonia galamensis. Lipid*, 33(12): 1217–1221.

Michaeli, W. and Kamps, T., 2007. Design of a micro injection moulding machine for thermosetting moulding materials: Assembly of hybrid microsystems. RWTH Aachen University, Aachen Germany, 4m–net.org.

Monteavaro, L.L., de Silva, E.O., Costa, A.P.O., Simons, D., Gerbase, A.E. and Petzhold, C.L., 2005. Polyurethane networks from formiated soy polyol: Synthesis and mechanical characterisation. *J. American Organic Chemistry Society*, 82(5): 365–371.

Neff, W.E., Konish, A.H. and Weisleder, D., 1993. High performance liquid chromatography of the triacylglycerols of *Vernonia galamensis* and Crepis alpine seed oils. *Ibid.*, 70: 449–453.

Prime, B.R., 1997. Thermosets. In: *Thermal Characterisation of Polymeric Materials, 2nd Edition*, (Ed.) E.A. Turi. Academic Press, London.

Purdy, R.H. and Craig, C.D., 1987. New source of long-chain fatty acids. *Ibid.* 64: 1493–1497.

Ravey, M. and Pearce, E.M., 1997. Flexible polyurethane foam. I. Thermal decomposition of a polyether-based, water-blown commercial type of flexible polyurethane foam. *J. Applied Polymer Science*, 63: 47–74.

Ren, B., Jiang, Y., Tan, K., Qin, S. and Zheng, X., 2007. Modification of euphorbia oil by ethoxylation and hydroxylation and its uses as fat liquoring agent. *Chemistry of Industrial Engineering Progress*, 26(5): 702–706.

Chapter 17

An Assessment of the Potential for Traditional and Modern Bamboo Construction Technologies in Uganda

Barnabas Nawangwe

PhD, Makerere University, Kampala, Uganda

ABSTRACT

Bamboo is internationally acknowledged as one of the most environmentally sustainable materials. It grows very fast, matures fast and has the best ability to clean the environment of carbon dioxide, while at the same time, it emits oxygen into the atmosphere very efficiently.

Bamboo grows and matures very fast and can be put to many uses for economic purposes, including construction of buildings and other structures. It can be used in both; traditional and modern technology in the form of pre-fabricated panels and beams. This paper highlights two demonstration bamboo construction projects at Makerere University in Uganda, the first using traditional and the second industrial bamboo construction technologies. The project aims to examine the qualities of industrial bamboo and its sustainability in the tropical conditions of Uganda. Preliminary results show that industrial bamboo has excellent environmental and strength qualities, pointing to the viability of the use of this material for solving environmental and socio-economic challenges that are faced by Uganda and countries with similar climatic and socio-economic conditions.

Keywords: *Bamboo, Environment, Industrialisation*

Introduction

According to Wikipedia (http://en.wikipedia.org/wiki/Bamboo), Bamboo is a group of perennial evergreens in the true grass family *Poaceae*, subfamily *Bambusoideae*, tribe *Bambuseae*.

Wikipedia also affirms, that bamboos are some of the fastest growing plants in the world, as some species are capable of growing 100 cm (39 inches.) or more per day due to a unique rhizome-dependent system. However, the growth rate is partially dependent on local soil and climatic conditions.

Bamboo is now recognised by many researchers as one of the most useful resources in the world. Together with several other sustainable and renewable resources, it is believed that bamboo can provide the potential to effectively replace wood in all of its various uses. Bamboo has been used for construction over many centuries in different parts of the world in using traditional technologies. 'In fully industrialized countries, bamboo will rarely stand on its own as the basic construction material, but can play a substantial role in quality construction, especially in light of the multitude of other materials to which we have access, many of which are similarly sustainable. In developing countries, however, where the citizenry presently inhabits simpler homes, bamboo may prove even more acceptable as a main building material. If it does, it can be used both to improve quality and to decrease cost of construction' (Edward Carney, 2011).

Li (2004, PhD thesis), *Physical, Chemical, And Mechanical Properties Of Bamboo And Its Utilization Potential For Fiberboard Manufacturing* affirms thus 'Over 1200 bamboo species have been identified globally. Bamboo has a very long history with human kind. Bamboo chips were used to record history in ancient China. Bamboo is also one of the oldest building materials used by human kind [Abd.Latif 1990]. It has been used widely for household products and extended to industrial applications due to advances in processing technology and increased market demand. In Asian countries, bamboo has been used for household utilities such as containers, chopsticks, woven mats, fishing poles, cricket boxes, handicrafts, chairs, etc. It has also been widely used in building applications, such as flooring, ceiling, walls, windows, doors, fences, housing roofs, trusses, rafters and purlins; it is also used in construction as structural materials for bridges, water transportation facilities and skyscraper scaffoldings'.

In the same study, Li (2004) examined the chemical and physical properties of bamboo that make it suitable construction, including specific gravity, compressive strength, moisture content and others. The study concludes that bamboo has very good strength properties with compressive strength of up to 85 Megapascal and Young's Modulus of up to 5000 MPa longitudinally for a 5-year bamboo stem. These strength properties demonstrate that bamboo is as strong as any other known structural construction material, including steel. Table 17.1 shows a comparison in strength properties between spruce, bamboo and steel.

'The replicability in Latin America and Africa of the success stories from South and South-east Asia is yet to be assessed, despite the immense interest from the private sector, non-governmental organizations and government institutions in using bamboo and rattan to fuel rural development in the region. The dearth of information on the bamboo and rattan sector has been the main constraint to the development of systematic and sustainable development programs in this sector.' (J. Francis Esegu, Joe Ssenteza and Jude Sekatuba, 2000).

Table 17.1: Comparison of Strength Properties for Spruce, Bamboo and Steel.

Type of Measurement	*Strength in kN/cm³*		
	Spruce	*Bamboo*	*Steel*
Elastic Modulus	110	2000	21000
Compressive strength	4.3	6.2-9.3	14
Tension	8,9	14,8-38,4	16
Bending	6,8	7,6-27,6	14
Shear	0,7	2,0	9,2

Source: Deutsche Bauzeitung (1997); accessed on 22nd July 2011 from file://D:\www.bambus\new\eng\reports\mechanical_properties\referat2.html.

Figure 17.1: Pedestrian Bridge in Bogota Built Using Traditional Bamboo Technology
***Source*: www.thearchitect.lk**

Methodology

This paper is based on a state of the art study of modern bamboo construction technologies around the world. This involved desk search for information about traditional and modern bamboo technologies in published literature and on the internet. Physical observation by the author was also utilized, especially when confirming bamboo technologies in Uganda. Laboratory tests were conducted on local bamboo species to confirm similarities in properties between data found in literature and the local bamboo. The paper bases its arguments, conclusions and recommendations on two bamboo demonstration projects at Makerere University, Uganda as case studies.

Bamboo Construction in Uganda

Uganda lies in the 'bamboo belt' of Africa. Figure 17.2 shows map of Africa with areas of considerable bamboo occurrence shaded green.

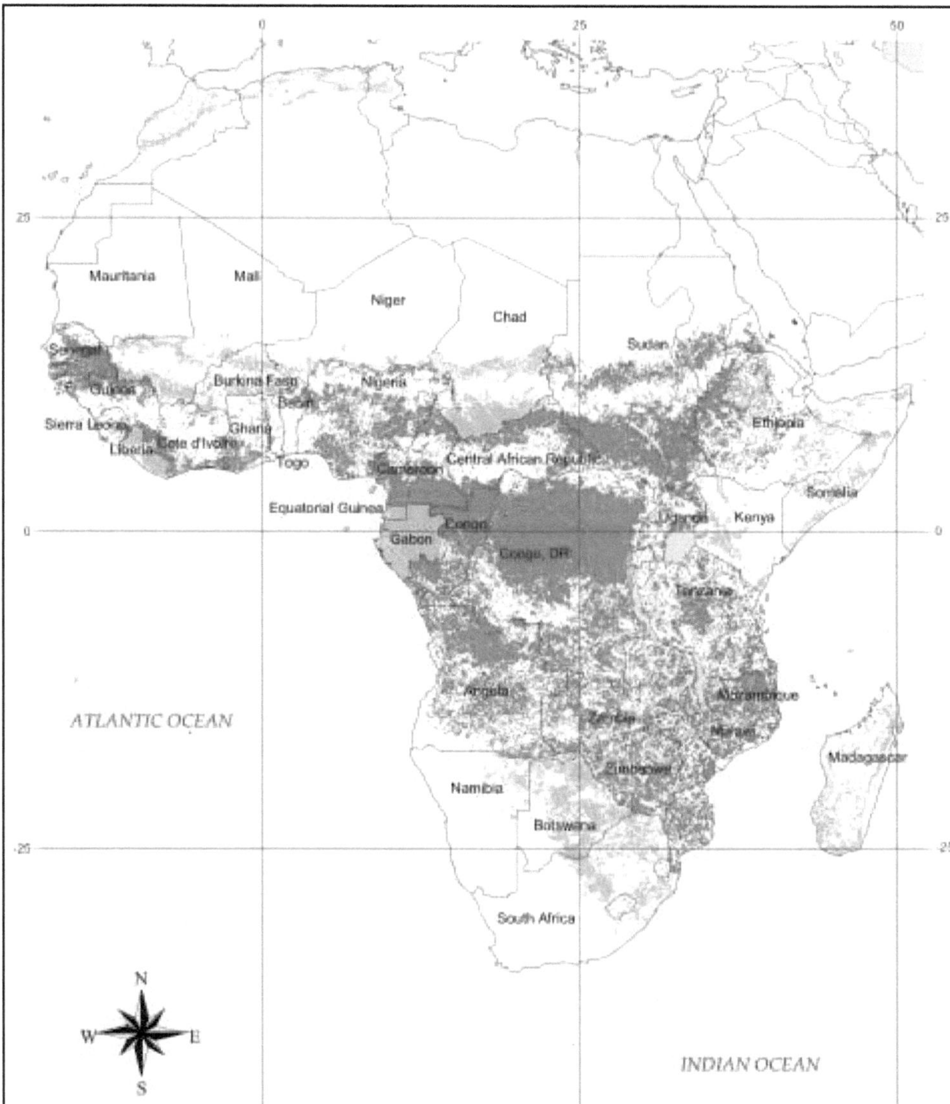

Figure 17.2: Map Showing Bamboo Distribution in Africa

***Source*: International Network for Bamboo and Rattan Beijing, China and UNEP World Conservation Monitoring Centre Cambridge**

'Rattan and bamboo resources in Uganda have traditionally been used by people as a source of building materials, for crafts, furniture, food (young bamboo shoots) and as well as areas for hunting wild game. The high productivity and economic potential of the rattan and bamboo sector in Uganda has barely been tapped...' (J.Francis Esegu, Joe Ssenteza and Jude Sekatuba,2000).

In Uganda, the bamboo most commonly used for construction is the mountain bamboo species, *Dendrocalmus*. This type of Bamboo is concentrated in the mountainous regions of the country in the east and the west.

'Bamboo harvesting and processing in Uganda is concentrated in the mountain forests of Elgon in Mbale District in the eastern part of Uganda, Rwenzori in the west and Echuya, Bwindi and Mgahinga in Kabale District in the South Western corner of the country. There are also other small pockets of bamboo vegetation scattered in various parts of Uganda.' (J.Francis Esegu, Joe Ssenteza and Jude Sekatuba,2000).

'Dry bamboo is used for the construction of houses, support for granaries, fencing, firewood, ceilings and making stretchers for carrying sick people. Ceilings and fencing are usually made under contract to the client. The contractor may employ a second person depending on the quantity and nature of work to be done. Granaries and beehives are made by weaving both green and dry bamboo. Despite this long tradition of using bamboo, there is widespread suspicion regarding the potential for its use in modern construction.

The conclusion below by a bamboo research group in India about perceptions of bamboo in India can very well be representative of the attitude to bamboo in Uganda:

Some of the key parameters are as below:

- ✰ *Image of bamboo*: One of the main problems to promote bamboo hosing is people's negative perception. In the most of the developing countries except a few countries in Latin America, people consider bamboo as a "poor men's timber" and living in a bamboo house carries a social stigma of being poor.
- ✰ *Scattered efforts*: There are independent efforts from various individuals and institutions to promote bamboo housing which is not enough to promote bamboo housing at global scale.
- ✰ *Resource availability*: Of more than 1200 species of bamboo, a limited number of bamboos are tested and approved for the construction. The mechanical properties of bamboo vary with species and location, therefore research and lab testing expensive and time consuming. This has significantly inhibited the development of a large scale bamboo housing project.
- ✰ *Proper use*: There is a serious threat that improper and non-scientific uses of bamboo for housing, and therefore quick deterioration of houses, may negatively affect the overall image of bamboo for building in the future. There is an urgent need to educate the public and building professionals at a global level and raise awareness to avoid such a threat.
- ✰ *Technical enhancement*: Bamboo is a non-dimensional material. Bamboo is hollow, not uniform in size and shape that make difficulties to use it as a substitute of timber. Bamboo requires technical enhancements and improved processing technology for its proper use.
- ✰ *Building code and standards*: Although, there is an international bamboo building code, the national building codes for bamboo are lacking in many countries. Lack of the codes and standards will limit its promotion especially to get building permission from the governments and to secure loans from the banks.

- *Technology*: No authenticated, proven and experimented bamboo housing technology exists in the country today. IIT (D) is the only institute which has experimented and demonstrated load bearing ability of bamboo housing.
- *New Concept*: The overall concept of bamboo is new in the country, till now bamboo was struck only as a craft item till NMBA came into picture and now slowly bamboo is being noticed.

Despite these challenges, the advantages of using bamboo in construction are enormous as confirmed by the results from the two demonstration buildings at Makerere University in Uganda.

Two demonstration buildings were erected, one built using traditional bamboo and the other pre-fabricated bamboo technology. The two buildings are shown in Figure 17.3.

Figure 17.3: Exterior and Interior of Traditional Bamboo Technology Houses and Pre-Fabricated Building at Makerere University

Laboratory tests were conducted, which confirmed that the chemical and physical properties of the local bamboo used were similar to results found in literature. A study of the two buildings at Makerere University confirmed the following advantages regarding use of bamboo for construction as suggested by Prashant (2008) :

- Bamboo is easy to cut, handle, repair, reposition and maintain, without the need for sophisticated tools or equipment.

- ☆ Because of its extraordinary physical characteristics, mountain bamboo is suitable for all types of structures and constructions.
- ☆ Bamboo is non-polluting and does not have crusts or parts that can be considered waste. Instead of adding to the problems of polluting land-fills like conventional building waste, any part of the bamboo that is not used is recycled back into the earth as fertilizer or can be processed as bamboo charcoal.
- ☆ Its circular form and hollow sections make bamboo a light building material, which is easy to handle, transport and store. Therefore, building with bamboo saves time.
- ☆ Bamboo can be utilized for permanent and for temporary constructions.
- ☆ In each of its nodes, bamboo has a dividing or transverse wall that maintains strength and allows bending thus preventing rupturing when bent. Because of this fantastic characteristic a bamboo construction offers superior earthquake-resistance.
- ☆ The composition of the fibers in the walls of the bamboo allows it to be cut length-wise or cross cut in pieces of any length, using simple manual tools like the machete.
- ☆ The natural surface of the bamboo is smooth, clean, with an attractive color which does not require painting, scraping or polishing.
- ☆ Besides being used as a structural element, bamboo can also serve other functions, such as: flooring, wall paneling, water pipes, drainage, and furniture.
- ☆ Bamboo can be used in combination with other types of construction materials, like reinforcing materials for foundations.

Table 17.2 adopted from Esegu, Ssenteza and Sekatuba (2000) summarises the status of bamboo use in Uganda. The table demonstrates the lack of awareness of the advantages of bamboo use, even for uses other than construction.

Conclusion and Recommendations

Bamboo is little used in the construction industry in Uganda. However, the potential for bamboo use is real and the advantages to environmental sustainability and socio-economic growth enormous. Uganda's climate is very good for bamboo farming, and there are already many farmers across the country engaged in it.

In order to encourage use of bamboo in construction, the following actions are recommended:

1. Conduct massive awareness campaigns about the looming dangers of deforestation and attendant lack of energy sources and building materials;
2. Encourage commercial bamboo farming through government subsidies, including provision of free bamboo seedlings to interested farmers;
3. Build many demonstration buildings around the country, especially in areas with high potential of bamboo propagation;

Table 17.2: Framework for Bamboo Processing

Sl.No.	*Activity*	*Findings*	*Causes*	*Constraints*	*Interventions*	*Opportunities*	*Output*
1.	Harvesting	☆ Bamboo harvested from natural stands	☆ No bamboo private plantations	☆ Permit to harvest must be obtained from FD and paid for by harvester	☆ Involvement of bamboo stake-holders in sustainable management ☆ Establish bamboo plantations	Collaborative approach by FORI and FD to forest management (CFM) FORI facilities	Improve sustainable management of bamboo supply raw materials of bamboo will be reliable
		☆ Tools used are quite rudimentary	☆ Lack of aware-ness about better techniques ☆ Lack of capital to buy better tools	Harvesting process is slow and laborious Bamboo sector is not developed	☆ Provision of modern harvesting technology to harvester	Awareness among the communities	☆ More efficient use of bamboo resources ☆ Better quality bamboo raw material ☆ Transporting to market centres simplified
		☆ Bamboo stocks are far away from the potential market	☆ There are no bamboo plantations ☆ No markets near bamboo sources	Raw materials trans-portation to potential market is hard and expensive .	☆ Establish primary/ secondary proce-ssing units near bamboo natural stands	☆ Good working relations between the FD and the communities	☆ Availability of Raw materials to potential market
2.	Processing	☆ Splitting is done manually	☆ Lack of modern processing tech-nology and techniques	☆ Inefficient processing ☆ Slow process ☆ Small amount of bamboo	Training processors in modern techni-ques Equipping of processors with tools	FORI facilities Existing groups	Availability of bamboo raw materials to manu-facturers Efficient process Foot Process More output

Contd...

Table 17.2–*Contd...*

Sl.No.	*Activity*	*Findings*	*Causes*	*Constraints*	*Interventions*	*Opportunities*	*Output*
3.	Manufacturing	✰ Few items produced from bamboo	✰ Lack of bamboo sector information by manufacturers ✰ Lack of training opportunities	✰ Low levels of craftsman skills	✰ Train manufacturers *i.e.* craftsmanship development ✰ Equipping manufacturers with current information on bamboo products/market development	Existing working groups can be strengthened Demand for bamboo products	Wider market will be captured ✰ Increased incomes of manufacturers ✰ Diversity of products
4.	Trading/ marketing	✰ Lack of marketing outlets	✰ Poor quality products ✰ Undeveloped sector	Low demand ✰ Small market	Craftsmanship development through training at manufacturing level	(See manufacturing)	
		✰ Items trades are few especially in Mbale	✰ Lack of bamboo sub-sector development	✰ Poor quality bamboo products ✰ Bamboo shoots poorly handled ✰ No preservation methods applied ✰ Lack of market developm ent	✰ Increase items produced from bamboo through craftsmanship development/ market development	✰ FORI facilities ✰ Involvement of women is vital in developing upon their craft weaving skills	Employment opportunities will be revised incomes of all stakeholders will be revised

4. Conduct research to identify fast-growing bamboos with favourable chemical and physical properties for construction purposes.

References

Bauzeitung, Deutsche, 1997. Comparison of Strength Properties for spruce, Bamboo and Steel. file: //D: \www.bambus\new\eng\reports\mechanical_properties\referat2.htmlBystriakova, N., Kapos, V. and Lysenko, I., 2004. International Network for Bamboo and Rattan Beijing, China and UNEP World Conservation Monitoring Centre Cambridge, UK.

Carney, Edward. Bamboo for Building the Future. http: //blog.greenearthbamboo.com/20110627/bamboo–worldwide–impact/bamboo–for–building–the–future/

Chihongo, A.W., Kishimbo, S.I. and Kachwele, M.D., 2000. Bamboo Production-to-Consumption Systems in Tanzania. INBAR Working Paper No. 28. International Network for Bamboo and Rattan, Beijing, China.

Esegu, J. Francis, Ssenteza, Joe and Sekatuba, Jude, 2000. Rattan and Bamboo in Uganda: A Study of the Production to Consumption Systems, International Network for Bamboo and Rattan, 2000.

Oscar Hidalgo Lopez, Manual de construccion con bambu, CIBAM, Universidad national de Columbia.

www.ingramcontent.com/pod-product-compliance
Ingram Content Group UK Ltd.
Pitfield, Milton Keynes, MK11 3LW, UK
UKHW040155091026
15587UKWH00018B/472

9 789351 242116